Deep Learning for Computer Vision with SAS®

An Introduction

Robert Blanchard

sas.com/books

Contents

About This Book

What Does This Book Cover?

Deep learning is an area of machine learning that has become ubiquitous with artificial intelligence. The complex, brain-like structure of deep learning models is used to find intricate patterns in large volumes of data. These models have heavily improved the performance of general supervised models, time series, speech recognition, object detection and classification, and sentiment analysis.

SAS has a rich set of established and unique capabilities with regard to deep learning. This book introduces the basics of deep learning with a focus on computer vision. The book details and demonstrates how to build computer vision models using SAS software. Both the "art" and science behind model building is covered.

Is This Book for You?

The general audience for this book should be either SAS or Python programmers with knowledge of traditional machine learning methods.

What Should You Know about the Examples?

This book includes tutorials for you to follow to gain hands-on experience with SAS.

Software Used to Develop the Book's Content

To follow along with the demos in this book, you will need the following software:

- SAS Viya (VDMML)

- SAS Studio

- Python

Example Code and Data

You can access the example code and data for this book by linking to its author page at https://support.sas.com/blanchard or on GitHub at https://github.com/sassoftware.

We Want to Hear from You

SAS Press books are written *by* SAS Users *for* SAS Users. We welcome your participation in their development and your feedback on SAS Press books that you are using. Please visit sas.com/books to do the following:

- Sign up to review a book
- Recommend a topic
- Request information on how to become a SAS Press author
- Provide feedback on a book

Do you have questions about a SAS Press book that you are reading? Contact the author through saspress@sas.com or https://support.sas.com/author_feedback.

SAS has many resources to help you find answers and expand your knowledge. If you need additional help, see our list of resources: sas.com/books.

Learn more about this author by visiting his author page https://support.sas.com/blanchard. There you can download free book excerpts, access example code and data, read the latest reviews, get updates, and more.

About The Author

Robert Blanchard is a Senior Data Scientist at SAS where he builds end-to-end artificial intelligence applications. He also researches, consults, and teaches machine learning with an emphasis on deep learning and computer vision for SAS. Robert has authored several professional courses on topics including neural networks, deep learning, and optimization modeling. Before joining SAS, Robert worked under the Senior Vice Provost at North Carolina State University, where he built models pertaining to student success, faculty development, and resource management. While working at North Carolina State University, Robert also started a private analytics company that focused on predicting future home sales. Prior to working in academia, Robert was a member of the research and development group on the Workforce Optimization team at Travelers Insurance. His models at Travelers focused on forecasting and optimizing resources. Robert graduated with a master's degree in Business Analytics and Project Management from the University of Connecticut and a master's degree in Applied and Resource Economics from East Carolina University.

Learn more about this author by visiting his author page https://support.sas.com/blanchard. There you can download free book excerpts, access example code and data, read the latest reviews, get updates, and more.

Chapter 1: Introduction to Deep Learning

Introduction to Neural Networks

Artificial neural networks mimic key aspects of the brain, in particular, the brain's ability to learn from experience. In order to understand artificial neural networks, we first must understand some key concepts of biological neural networks, in other words, our own biological brains.

A biological brain has many features that would be desirable in artificial systems, such as the ability to learn or adapt easily to new environments. For example, imagine you arrive at a city in a country that you have never visited. You don't know the culture or the language. Given enough time, you will learn the culture and familiarize yourself with the language. You will know the location of streets, restaurants, and museums.

The brain is also highly parallel and therefore very fast. It is not equivalent to one processor, but instead it is equivalent to a multitude of millions of processors, all running in parallel. Biological brains can also deal with information that is fuzzy, probabilistic, noisy, or inconsistent, all while being robust, fault tolerant, and relatively small. Although inspired by cognitive science (in particular, neurophysiology), neural networks largely draw their methods from statistical physics (Hertz et al. 1991). There are dozens, if not hundreds, of neural network algorithms.

Biological Neurons

In order to imitate neurons in artificial systems, first their mechanisms needed to be understood. There is still much to be learned, but the key functional aspects of neurons, and even small systems (networks) of neurons, are now known.

Neurons are the fundamental units of cognition, and they are responsible for sending information from the brain to the rest of the body. Neurons have three parts: a cell body, dendrites, and axons. Inputs arrive in the dendrites (short branched structures) and are transmitted to the next neuron in the chain via the axons (a long, thin fiber). Neurons do not actually touch each other but communicate across the gap (called a synaptic gap) using neurotransmitters. These chemicals either excite the receiving neuron, making it more likely to "fire," or they inhibit the neuron, making it less likely to become active. The amount of neurotransmitter released across the gap determines the relative strength of each dendrite's connection to the receiving neuron. In essence, each synapse "weights" the relative strength of its arriving input. The synaptically weighted inputs are summed. If the sum exceeds an adaptable threshold (or bias) value, the neuron sends a pulse down its axon to the other neurons in the network to which it connects.

A key discovery of modern neurophysiology is that synaptic connections are adaptable; they change with experience. The more active the synapse, the stronger the connection becomes. Conversely, synapses with little or no activity fade and, eventually, die off (atrophy). This is thought to be the basis of learning. For example, a study from the University of Wisconsin in 2015 showed that people could begin to "see" with their tongue. Attached to the electric grid was a camera that was fastened to the subject's forehead. The subject was blindfolded. However, within 30 minutes, as their neurons adapted, subjects began to "see" with their tongue. Amazing!

Although there are branches of neural network research that attempt to mimic the underlying biological processes in detail, most neural networks do not try to be biologically realistic.

Mathematical Neurons

In a seminal paper with the rather understated title "A logical calculus of the ideas immanent in nervous activity," McCulloch and Pitts (1943) gave birth to the field of artificial neural networks. The fundamental element of a McCulloch-Pitts network is called, unsurprisingly, a McCulloch-Pitts neuron. As in real neurons, each input (x_i) is first weighted (w_i) and then summed. To mimic a neuron's threshold functionality, a bias value (w_0) is added to the weighted sum, predisposing the neuron to either a positive or negative output value. The result is known as the neuron's net input:

$$net = w_0 + \sum_{i=1}^{k} w_i x_i$$

Notice that this is the classic linear regression equation, where the bias term is the y-intercept and the weight associated with each input is the input's slope parameter.

The original McCulloch-Pitts neuron's final output was determined by passing its net input value through a step function (a function that converts a continuous value into a binary output 0 or 1, or a bipolar output -1 or 1), turning each neuron into a linear classifier/discriminator. Modern neurons replace the discontinuous step function used in the McCulloch-Pitts neuron with a continuous function. The continuous nature permits the use of derivatives to explore the parameter space.

$$H = f\left(w_0 + \sum_{i=1}^{d} w_i x_i\right)$$

The mathematical neuron is considered the cornerstone of a neural network. There are three layers in the basic multilayer perceptron (MLP) neural network:

1. An *input layer* containing a neuron/unit for each input variable. The input layer neurons have no adjustable parameters (weights). They simply pass the positive or negative input to the next layer.
2. A *hidden layer* with hidden units (mathematical neurons) that perform a nonlinear transformation of the weighted and summed input activations.
3. An *output layer* that shapes and combines the nonlinear hidden layer activation values.

A single hidden-layer multilayer perceptron constructs a limited extent region, or *bump*, of large values surrounded by smaller values (Principe et al. 2000). The intersection of the hyper-planes created by a hidden layer consisting of three hidden units, for example, forms a triangle-shaped bump.

The hidden and output layers ***must not*** be connected by a strictly linear function in order to act as separate layers. Otherwise, the multilayer perceptron collapses into a linear perceptron. More formally, if matrix **A** is the set of weights that transforms input matrix **X** into the hidden layer output values, and matrix **B** is the set of weights that transforms the hidden unit output into the final estimates **Y**, then the linearly connected multilayer network can be represented as **Y=B[A(X)]**. However, if a single-layer weight matrix **C=BA** is created, exactly the same output can be obtained from the single-layer network—that is, **Y=C(X)**.

In a two-layer perceptron with k inputs, h_1 hidden units in the first hidden layer, and h_2 hidden units in the second hidden layer, the number of parameters to be learned is $h_1(k+1) + h_2(h_1+1) = h_2 = 1$.

The number 1 represents the biased weight W_0 in the combination function of each neuron.

Figure 1.1: Multilayer Perceptron

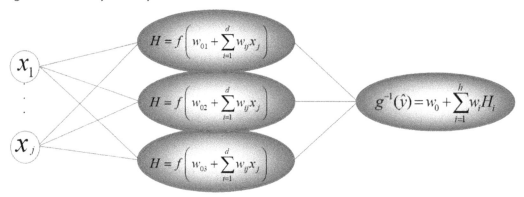

> **Note:** The "number of parameters" equations in this book assume that the inputs are interval or ratio level. Each nominal or ordinal input increases *k* by the number of classes in the variable, minus 1.

Deep Learning

The term *deep learning* refers to the numerous hidden layers used in a neural network. However, *the true essence of deep learning is the methods that enable the increased extraction of information* derived from a neural network with more than one hidden layer. Adding more hidden layers to a neural network provides little benefit without deep learning methods that underpin the efficient extraction of information. For example, SAS software has had the capability to build neural networks with many hidden layers using the NEURAL procedure for several decades. However, a case can be made to suggest that SAS has not had deep learning because the key elements that enable learning to persist in the presence of many hidden layers had not been discovered. These elements include the use of the following:

- activation functions that are more resistant to saturation than conventional activation functions
- fast moving gradient-based optimizations such as Stochastic Gradient Descent and ADAM
- weight initializations that consider the amount of incoming information
- new regularization techniques such as dropout and batch normalization
- innovations in distributed computing.

The elements outlined above are included in today's SAS software and are described below. Needless to say, deep learning has shown impressive promise in solving problems that were previously considered infeasible to solve.

The process of deep learning is to formulate an outcome from engineering new glimpses of the input space, and then reengineering these engineered projections with the next hidden layer. This process is repeated for each hidden layer until the output layers are reached. The output layers reconcile the final layer of incoming hidden unit information to produce a set of outputs. The classic example of this process is facial recognition. The first hidden layer captures shades of the image. The next hidden layer combines the shades to formulate edges. The next hidden layer combines these edges to create projections of ears, mouths, noses, and other distinct aspects that define a human face. The next layer combines these distinct formulations to create a projection of a more complete human face. And so on. A brief comparison of traditional neural networks and deep learning is shown in Table 1.1.

Table 1.1: Traditional Neural Networks versus Deep Learning

Aspect	Traditional	Deep Learning
Hidden activation function(s)	Hyperbolic Tangent (tanh)	Rectified Linear (ReLU) and other variants
Gradient-based learning	Batch GD and BFGS	Stochastic GD, Adam, and LBFGS
Weight initialization	Constant Variance	Normalized Variance
Regularization	Early Stopping, L1, and L2	Early Stopping, L1, L2, Dropout, and Batch Normalization
Processor	CPU	CPU or GPU

Deep learning incorporates activation functions that are more resistant to neuron saturation than conventional activation functions. One of the classic characteristics of traditional neural networks was the infamous use of sigmoidal transformations in hidden units. Sigmoidal transformations are problematic for gradient-based learning because the sigmoid has two asymptotic regions that can saturate (that is, gradient of the output is near zero). The red or deeper shaded outer areas represent areas of saturation. See Figure 1.2.

Figure 1.2: Hyperbolic Tangent Function

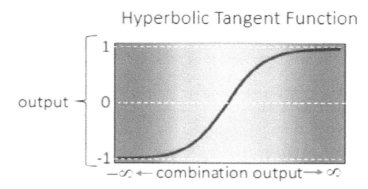

On the other hand, a linear transformation such as an identity poses little issue for gradient-based learning because the gradient is a constant. However, the use of linear transformations negates the benefits provided by nonlinear transformations (that is, approximate nonlinear relationships).

Rectified linear transformation (or ReLU) consists of piecewise linear transformations that, when combined, can approximate nonlinear functions. (See Figure 1.3.)

Figure 1.3: Rectified Linear Function

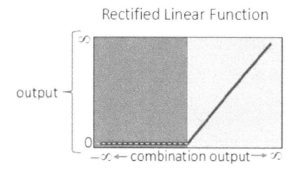

In the case of ReLU, the derivative for the ***active*** region output by the transformation is 1 and 0 for the ***inactive*** region. The inactive region of the ReLU transformation can be viewed as a weakness of the transformation because it inhibits the unit from contributing to gradient-based learning.

The saturation of ReLU could be somewhat mitigated by cleverly initializing the weights to avoid negative output values. For example, consider a business scenario of modeling image data. Each unstandardized input pixel value ranges between 0 and 255. In this case, the weights could be initialized and constrained to be strictly positive to avoid negative output values, avoiding the non-active output region of the ReLU.

Other variants of the rectified linear transformation exist that permit learning to continue when the combination function resolves to a negative value. Most notable of these is the exponential linear activation transformation (ELU) as shown in Figure 1.4.

Figure 1.4: Exponential Linear Function

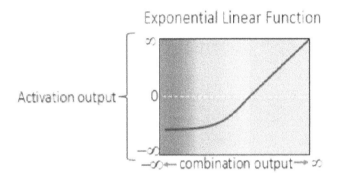

SAS researchers have observed better performance when ELU is used instead of ReLU in convolutional neural networks in some cases. SAS includes other, popular activation functions that are not shown here, such as softplus and leaky. Additionally, you can create your own activation functions in SAS using the SAS Function Compiler (or FCMP).

> **Note:** Convolutional neural networks (CNNs) are a class of artificial neural networks. CNNs are widely used in image recognition and classification. Like regular neural networks, a CNN consists of multiple layers and a number of neurons. CNNs are well suited for image data, but they can also be used for other problems such as natural language processing. CNNs are detailed in Chapter 2.

The error function defines a surface in the parameter space. If it is a linear model fit by least squares, the error surface is convex with a unique minimum. However, in a nonlinear model, this error surface is often a complex landscape consisting of numerous deep valleys, steep cliffs, and long-reaching plateaus.

To efficiently search this landscape for an error minimum, optimization must be used. The optimization methods use local features of the error surface to guide their descent. Specifically, the parameters associated with a given error minimum are located using the following procedure:

1. Initialize the weight vector to small random values, $\mathbf{w}^{(0)}$.
2. Use an optimization method to determine the update vector, $\delta^{(t)}$.

3. Add the update vector to the weight values from the previous iteration to generate new estimates:

$$\mathbf{w}^{(t+1)} = \mathbf{w}^{(t)} + \delta^{(t)}$$

4. If none of the specified convergence criteria have been achieved, then go back to step 2.

Here are the three conditions under which convergence is declared:

1. when the specified error function stops improving
2. if the gradient has no slope (implying that a minimum has been reached)
3. if the magnitude of the parameters stops changing substantially

Batch Gradient Descent

Re-invented several times, the back propagation (*backprop*) algorithm initially just used *gradient descent* to determine an appropriate set of weights. The gradient, $\nabla \mathbf{g}^{(t)}$, is the vector of partial derivatives of the error function with respect to the weights. It points in the steepest direction uphill. (See Figure 1.5.)

Figure 1.5: Batch Gradient Descent

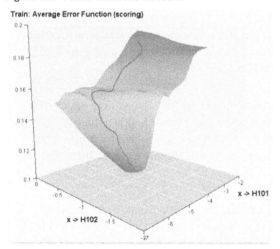

By negating the step size (that is, *learning rate*) parameter, η, a step is made in the direction that is locally steepest downhill:

$$\delta^{(t)} = -\eta \nabla \mathbf{g}^{(t)}$$

The parameters associated with a given error minimum are located using the following procedure:

1. Initialize the weight vector to small random values, $\mathbf{w}^{(0)}$.
2. Use an optimization method to determine the update vector, $\delta^{(t)}$.
3. Add the update vector to the weight values from the previous iteration to generate new estimates:

$$\mathbf{w}^{(t+1)} = \mathbf{w}^{(t)} + \delta^{(t)}$$

5. If none of the specified convergence criteria has been achieved, then back go to step 2.

Unfortunately, as gradient descent approaches the desired weights, it exhibits numerous back-and-forth movements known as *hemstitching*. To control the training iterations wasted in this hemstitching, later versions of back propagation included a momentum term, yielding the modern update rule:

$$\delta^{(t)} = -\eta \nabla \mathbf{g}^{(t)} + \alpha \delta^{(t-1)}$$

The momentum term retains the last update vector, $\delta^{(t-1)}$, using this information to "dampen" potentially oscillating search paths. The cost is an extra learning rate parameter ($0 \le \alpha \le 1$) that must be set. This updated rule uses all the training observations (t) to calculate the exact gradient on each descent step. This results in a smooth progression to the gradient minima.

Stochastic Gradient Descent

In the batch variant of the gradient descent algorithm, generation of the weight update vector is determined by using all of the examples in the training set. That is, the exact gradient is calculated, ensuring a relatively smooth progression to the error minima.

However, when the training data set is large, computing the exact gradient is computationally expensive. The entire training data set must be assessed on each step down the gradient. Moreover, if the data are redundant, the error gradient on the second half of the data will be almost identical to the gradient on the first half. In this event, it would be a waste of time to compute the gradient on the whole data set. You would be better off computing the gradient on a subset of the weights, updating the weights, and then repeating on a new subset. In this case, each weight update is based on an approximation to the true gradient. But as long as it points in approximately the same direction as the exact gradient, the approximate gradient is a useful alternative to computing the exact gradient (Hinton 2007).

Taken to extremes, calculation of the approximate gradient can be based on a single training case. The weights are then updated, and the gradient is calculated on the next case. This is known as *stochastic gradient descent* (also known as *online learning*). (See Figure 1.6.)

Figure 1.6: Stochastic Gradient Descent

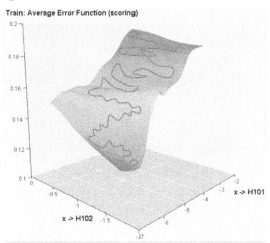

Stochastic gradient descent is very effective, particularly when combined with a momentum term, $\delta^{(t-1)}$:

$$\delta^{(t)} = -\eta \nabla \mathbf{g}^{(t)} + \alpha \delta^{(t-1)}$$

Because stochastic gradient descent does not need to consider the entire training data set when calculating each descent step's gradient, it is usually faster than batch gradient descent. However, because each iteration is trying to better fit a single observation, some of the gradients might actually point away from the minima. This means that, although stochastic gradient descent generally moves the parameters in the direction of an error minima, it might not do so on each iteration. The result is a more circuitous path. In fact, stochastic gradient descent does not actually converge in the same sense as batch gradient descent does. Instead, it wanders around continuously in some region that is close to the minima (Ng, 2013).

Introduction to ADAM Optimization

The ADAM method applies adjustments to the learned gradients for each individual model parameter in an adaptive manner by approximating second-order information about the objective function based on previously observed mini-batch gradients. The "adaptive movement" nature of the algorithm's movement is where the name ADAM comes from (Kingma and Ba, 2014).

The ADAM method introduces two new hyperparameters to the mix, (β_1^t) and (β_2^t) where t represents the iteration count. A learning rate that controls the originating step size is also included. The adjustable beta terms are used to approximate a *signal-to-noise* ratio that is used to scale the step size. When the approximated single-to-noise ratio is large, the step size is closer to the originating step size (that of traditional stochastic gradient descent).

When the approximated single-to-noise ratio is small, the step size is near zero. This is a nice feature because a lower single-to-noise ratio is an indication of higher uncertainty. Thus, more cautious steps should be taken in the parameter space (Kingma and Ba 2014).

To use ADAM, specify 'ADAM' in the METHOD= suboption of the ALGORITHM= option in the OPTIMIZER parameter. The suboptions for β_1 and β_2, as well as the α and other options, also need to be specified. In the example code below, β_1 = .9, β_2 = .999 and α = .001.

```
optimizer={algorithm={method='ADAM',
                      beta1=0.9,
                      beta2=0.999,
                      learningrate=.001,
                      lrpolicy='Step',
                      gamma=0.5},
                      minibatchsize=100,
                      maxepochs=200}
```

> **Note:** The authors of ADAM recommend a β_1 value of .9, a β_2 value of .999, and an α (learning rate) of .001.

Weight Initialization

Deep learning uses different methods of weight initialization than traditional neural networks do. In neural networks, the hidden unit weights are randomly initialized to ensure that each hidden unit is approximating different areas of relationship between the inputs and the output. Otherwise, each hidden unit would be approximating the same relational variations if the weights across hidden units were identical, or even symmetric. The hidden unit weights are usually randomly initialized to some specified distribution, commonly Gaussian or Uniform.

Traditional neural networks use a standard variance for the randomly initialized hidden unit weights. This can become problematic when there is a large amount of incoming information (that is, a large number of incoming connections) because the variance of the hidden unit will likely increase as the amount of incoming connections increases. This means that the output of the combination function *could* be more extreme, resulting in a saturated hidden unit (Daniely et al. 2017).

Deep learning methods use a normalized initialization in which the variance of the hidden weights is a function of the amount of incoming information and outgoing information. SAS offers several methods for reducing the variance of the hidden weights. *Xavier* initialization is one of the most common weight initialization methods used in deep learning. The initialization method is random uniform with variance

$$w_{i,j} \sim U(-\sqrt{\frac{6}{m+n}}, \sqrt{\frac{6}{m+n}})$$

where *m* is the number of input connections (fan-in) and *n* is the number of output connections (fan-out) (hidden units in current layer).

One potential flaw of the Xavier initialization is that the initialization method assumes a linear activation function, which is typically not the case in hidden units. *MSRA* was designed with the ReLU activation function in mind because MSRA operates under the assumption of a nonzero mean output by the activation, which is exhibited by ReLU (He et al. 2015). The MSRA initialization method is random Gaussian distribution with a standardization of

$$\sqrt{\frac{2}{avg\,(m+n)}}$$

SAS includes a second variant of the MSRA, called *MSRA2*. Similar to the MSRA initialization, the MSRA2 method is a random Gaussian distribution with a standardization of

$$\sqrt{\frac{2}{n}}$$

And it penalizes only for outgoing (fan-out) information.

> **Note:** Weight initializations have less impact over model performance if batch normalization is used because batch normalization standardizes information passed between hidden layers. Batch normalization is discussed later in this chapter.

Consider the following simple example where unit y is being derived from 25 randomly initialized weights. The variance of unit y is larger when the standard deviation is held constant at 1. This means that the values for y are more likely to venture into a saturation region when a nonlinear activation function is incorporated. On the other hand, Xavier's initialization penalizes the variance for the incoming and outgoing connections, constraining the value of y to less treacherous regions of the activation. See Figures 1.7 and 1.8, noting that these examples assume that there are 25 incoming and outgoing connections.

Figure 1.7: Constant Variance (Standard Deviation = 1)

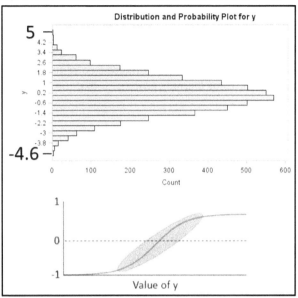

Figure 1.8: Constant Variance (Standard Deviation = $\sqrt{\dfrac{6}{25+25}} \approx .34$**)**

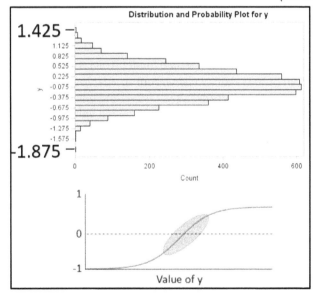

Regularization

Regularization is a process of introducing or removing information to stabilize an algorithm's understanding of data. Regularizations such as early stopping, L1, and L2 have been used extensively in neural networks for many years. These regularizations are still widely used in deep learning, as well. However, there have been advancements in the area of regularization that work particularly well when combined with multi-hidden layer neural networks. Two of these advancements, dropout and batch normalization, have shown significant promise in deep learning models. Let's begin with a discussion of dropout and then examine batch normalization.

Dropout adds noise to the learning process so that the model is more generalizable. Training an ensemble of deep neural networks with several hundred thousand parameters each might be infeasible. As seen in Figure 1.9, dropout adds noise to the learning process so that the model is more generalizable.

Figure 1.9: Regularization Techniques

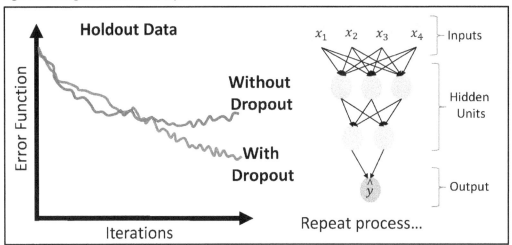

The goal of dropout is to approximate an ensemble of many possible model structures through a process that perturbs the learning in an attempt to prevent weights from co-adapting. For example, imagine we are training a neural network to identify human faces, and one of the hidden units used in the model sufficiently captures the mouth. All other hidden units are now relying, at least in some part, on this hidden unit to help identify a face through the presence of the mouth. Removing the hidden unit that captures the mouth forces the remaining hidden units to adjust and compensate. This process pushes each hidden unit to be more of a "generalist" than a "specialist" because each hidden unit must reduce its reliance on other hidden units in the model.

During the process of dropout, hidden units or inputs (or both) are randomly removed from training for a period of weight updates. Removing the hidden unit from the model is as simple as multiplying the unit's output by zero. The removed unit's weights are not lost but rather frozen. Each time that units are removed, the resulting network is referred to as a *thinned network*. After several weight updates, all hidden and input units are returned to the network. Afterward, a new subset of hidden or input units (or both) are randomly selected and removed for several weight updates. The process is repeated until the maximum training iterations are reached or the optimization procedure converges.

In SAS Viya, you can specify the DROPOUT= option in an ADDLAYER statement to implement dropout. DROPOUT=*ratio* specifies the dropout ratio of the layer.

Below is an example of dropout implementation in an ADDLAYER statement.

```
AddLayer/model='DLNN' name="HLayer1" layer={type='FULLCONNECT' n=30
        act='ELU' init='xavier' dropout=.05} srcLayers={"data"};
```

> **Note:** The ADDLAYER syntax is described shortly and further expanded upon throughout this book.

Batch Normalization

The *batch normalization* (Ioffe and Szegedy, 2015) operation normalizes information passed between hidden layers per mini-batch by performing a standardizing calculation to each piece of input data. The standardizing calculation subtracts the mean of the data and then divides by the standard deviation. It then follows this calculation by multiplying the data by the value of a learned constant and then adding the value of another learned constant.

Thus, the normalization formula is

$$\gamma * (\frac{X_i - \mu}{\sigma}) + \beta$$

where gamma (γ) and beta (β) are learnable parameters.

Some deep learning practitioners have dismissed the use of sigmoidal activations in the hidden units. Their dismissal might have been premature, however, with the discovery of batch normalization. Without batch normalization, each hidden layer is, in essence, learning from information that is constantly changing when multiple hidden layers are present in a neural network. That is, a weight update is reliant on second-order, third-order (and so on) effects (weights in the other layers). This phenomenon is known as the *internal covariance shift* (ICS) (Ioffe and Szegedy, 2015).

There are two schools of thought as to why batch normalization improves the learning process. The first comes from Ioffe and Szegedy who believe batch normalization reduces ICS. The second comes from Santurkar, Tsipras, Ilyas, and Madry who argue that batch normalization is not really reducing ICS but is instead smoothing the error landscape (Santurkar, Tsipras, Ilyas, and Madry 2018). Regardless of which thought prevails, batch normalization has empirically shown to improve the learning process and reduce neuron saturation.

In the SAS deep learning actions, batch normalization is implemented as a separate layer type and can be placed anywhere after the input layer and before the output layer.

> **Note:** With regard to convolutional neural networks, the batch normalization layer is typically inserted after a convolution or pooling layer.

Batch Normalization with Mini-Batches

In the case where the source layer to a batch normalization layer contains feature maps, the batch normalization layer computes statistics based on all of the pixels in each feature map, over all of the observations in a mini-batch. For example, suppose that your network is configured for a mini-batch size of 3, and the input to the batch normalization layer consists of two 5 x 5 feature maps. In this case, the batch normalization layer computes two means and two standard deviations. The first mean would be the mean of all the pixels in the first feature map for the first observation, the first feature map of the second observation, and the first feature map of the third observation. The second mean would be the mean of all of the pixels in the second feature map of the first observation, the second feature map of the second observation, and the second feature map of the third observation, and so on. Numerically, each mean would be the mean of (3 x 5 x 5) = 75 values.

In the case where the source layer to a batch normalization layer does not contain feature maps (for example, a fully connected layer), then the batch normalization layer computes statistics for each neuron in the input, rather than for each feature map in the input. For example, suppose that your network has a mini-batch size of 3, and the input to the batch normalization layer contains 50 neurons. In this case, the batch normalization layer would compute 50 means and 50 standard deviations. The first mean would be the mean of the first neuron of the first observation, the first neuron of the second observation, and the first neuron of the third observation. The second mean would be the mean of the second neuron of the first observation, the second neuron of the second observation, and the second neuron of the third observation, and so on. Numerically, each mean would be the mean of three values. NVIDIA refers to this calculation as *per activation* mode.

In order for the batch normalization computations to conform to those described in Sergey Ioffe and Christian Szegedy's batch normalization research (Ioffe and Szegedy, 2015), the source layer should have settings of ACT=IDENTITY and INCLUDEBIAS=FALSE. The activation function that would normally have been specified in the source layer should instead be specified on the batch normalization layer. If you do not configure your model to follow these option settings, the computation will still work, but it will not match the computation as described by Ioffe and Szegedy.

When using multiple GPUs, efficient calculation of the batch normalization transform requires a modification to the original algorithm specified by Ioffe and Szegedy. The algorithm specifies that during training, you must calculate the mean and standard deviation of the pixel values in each feature map, over all of the observations in a mini-batch.

However, when using multiple GPUs, the observations in the mini-batch are distributed over the GPUs. It would be very inefficient to try to synchronize each GPU's batch normalization calculations for each batch normalization layer. Instead, each GPU calculates the required statistics using a subset of available observations and uses those statistics to perform the transformation on those observations.

Research communities are still debating whether small or large minibatch sizes yield better performance. However, when a minibatch of observations is distributed across multiple GPUs, and the model contains batch normalization layers, the deep learning team at SAS recommends that you use reasonably large-sized mini-batches on each GPU so that the statistics will be stable.

In addition to calculating feature map statistics on each mini-batch, the batch normalization algorithm also needs to calculate statistics over the entire training data set before saving the training weights. These statistics are the ones used for scoring (whereas the mini-batch statistics are used for training). Rather than perform an extra epoch at the end of training, the statistics from each mini-batch are averaged over the course of the last training epoch to create the epoch statistics.

The statistics computed in this way are a close approximation to the more complicated computation that uses an extra epoch with fixed weights (as long as the weights in the last epoch do not change much) after each mini-batch of the epoch. (This is usually the case for the last training epoch.) When using multiple GPUs, this calculation is performed exactly the same way as when using a single GPU. That is, the statistics for each mini-batch on each GPU are averaged after each mini-batch to compute the final epoch statistics for scoring.

Traditional Neural Networks versus Deep Learning

Recall the differences between traditional neural networks and deep learning are shown in Table 1.2. Traditional neural networks leveraged the computation of a single central processing unit (CPU) to train the model. However, graphical processing units (GPUs) have a design that naturally fits well with the structure and learning process of neural networks. There have been promising developments in the use of CPUs grouped together that use a fixed-point architecture as opposed to a floating-point architecture (Vanhoucke et al. 2011). The details of the distribution of computation is a deeply complex topic and remains outside the scope of this book, although this brief comparison of CPUs to GPUs is provided in Table 1.2.

Table 1.2: Comparison of Central Processing Units and Graphical Processing Units

Central Processing Unit (CPU)	Graphical Processing Unit (GPU)
Faster Clock Speed	Slower Clock Speed
Fewer Processing Units	More Processing Units
More Branching	Less Branching
Less Memory Bandwidth	More Memory Bandwidth

The optimization techniques used to adjust the weights of a neural network are iterative processes. However, within each iteration, the weights are updated simultaneously. Therefore, calculations corresponding to each weight update can be distributed among processing units.

GPUs are designed to perform many operations in parallel, which fits nicely with the weight update process used by neural networks.

The use of GPUs should be reserved for larger neural networks because the difference in performance between CPUs and GPUs is negligible in neural networks with a small number of parameters.

Deep Learning Actions

As an integrated part of the SAS Platform, SAS Viya is a cloud-enabled, in-memory analytics engine that provides quick, accurate, and reliable analytical insights. SAS Viya offers a rich set of data mining and machine learning capabilities that run on a robust in-memory distributed computing infrastructure that provides a single environment that is unified, open, powerful, and cloud ready.

The SAS Cloud Analytic Services actions can be surfaced through SAS Viya on a number of interfaces, including SAS Studio and Jupyter notebook.

This book highlights three of the deep learning actions in SAS Cloud Analytic Services (CAS):

- deep feed-forward neural network (DNN)
- convolutional neural network (CNN)
- recurrent neural network (RNN)

DNN actions are used to solve more traditional classification problems, such as fraud detection. CNN actions are commonly used to build more advanced neural networks for either traditional or computer vision data problems. An RNN is used to solve problems for data that is some function of a sequence, such as time series or text analyses.

SAS deep learning actions can be called using several programming languages, including SAS, R, and Python. This book focuses on the use of SAS to call Cloud Analytic Services through the CAS procedure.

The CAS procedure enables you to interact with SAS Cloud Analytic Services from the SAS client by providing a programming environment based on the CASL language specification. The programming environment enables you to run CAS actions and use the results to prepare the parameters for another action. Code is formatted as

```
PROC CAS;
     <CASL code>
Quit;
```

An example of this is

```
PROC CAS < exc >< noqueue >;
BuildModel/ modeltable={name="<Model table name >"}
           type="DNN";
Quit;
```

For CNNs and RNNs, replace the type="DNN" with type="CNN" and type="RNN", respectively.

The CAS procedure has several features that enable you to perform the following operations:

- run any CAS action that is supported by the server, even if the action did not exist at the time of the release
- use multiple sessions to perform asynchronous execution
- operate on parameters and results as variables using the full function expression parser
- import your own executables that define callable functions

Optional arguments such as those below can be used in the PROC CAS statement.

- EXC: Executes the CASL code as soon as the previous block of code has completed processing. The default option does not execute CASL code until a RUN statement is entered.
- NOQUEUE: Forces output to be displayed as soon as output is produced.

> **Note:** Global statements, SAS macro code, and RUN statements do not terminate the CAS procedure.

Building a Deep Neural Network

Deep neural networks can be created using the following code:

```
PROC CAS < exc ><noqueue >;
BuildModel/ modeltable={name="<model table name >"} type ="DNN";
AddLayer/
      modeltable="<model table name >"
      name="<name of layer >"
      layer={type="layer type"
         n="< number of hidden units >"
         act="< type of activation transformation >"
         init="< weight initialization method >"}
      srcLayers={"< previous layer name >"};
Quit;
```

The AddLayer action is used to add a layer to a deep learning neural network. Table 1.3 provides examples of different types of layers that are used throughout the book.

Table 1.3: Layer Types

Type	Details	Example
INPUT	Input Layer	layer={type='input'}
CONVO/CONVOLUTION	Convolutional Layer	layer={type='convolution' nFilters=32 width=5 height=5 stride=1 act='tanh' }
POOLING	Pooling Layer	layer={type='pooling' width=2 height=2 stride=2}
FC/FULLCONNECT	Fully connected Layer	layer={type='fullconnect' n=50 act='sigmoid' }
OUTPUT	Output Layer	layer={type='output' act='softmax' }
RESIDUAL	Residual Layer	layer={type='residual' }
BATCHNORM	Batch Normalization	layer={type='batchnorm' act='ELU'}
CONCAT	Concatenation Layer	layer={type='concat'}
FCMP	FCMP Layer	layer={type='FCMP', forwardFunc='forward_prop', backwardFunc='back_prop', height=1, width=40, depth=1, nweights=1280}
RECURRENT	Recurrent Layer	layer={type='recurrent' n=50 act='sigmoid' rnnType='gru'}
TRANSCONVO	Transpose Convolution Layer	layer={type='transconvon' Filters=32 width=5 height=5 stride=2 act='tanh' }

> **Note:** Many other layers are available but not described in this book including: DETECTION, PROJECTION, FCMP, CLOSS, NASLAYER, SPLIT, CHANNELSHUFFLE, CLUSTER, FASTRCNN, GROUPCONVO, LAYERNORM, MASKRCNN, MHATTENTION, REGIONPROPOSAL, ROIALIGN, ROIPOOLING, SCALE, SEGMENTATION, SURVIVAL, and many others. The DETECTION layer is used for object detection, and the FCMP layer is used to implement user-designed activation and error functions. The CLUSTER layer is used for DeepCluster and should be connected to a fully connected layer. The other layers are described in detail in SAS documentation.

Training a Deep Learning CAS Action Model

The dlTrain action can be used to train a deep learning model. The parameters of the dlTrain action can be found in SAS documentation by searching for "Deep Learning Action Set: Syntax". Here is some example code with parameters:

```
PROC CAS;
   dlTrain /
      table=' CAS-libref.data-table '
      modeltable= ' model name, specified in buildmodel action '
      bestweights={name= 'Name of output table containing best
                                       model weights' }
      inputs={' List of ALL input variables '}
      nominals={' List of nominal input variables '}
      validtable={name= 'Name of validation data table' }
      target={' list of target variables'};
Quit;
```

> **Note:** When training your model, you can use the OPTIMIZER parameter to specify the settings for the optimization algorithm, optimization mode, and other settings such as a seed, the maximum number of epochs, and so on.

Demonstration 1: Loading and Modeling Data with Traditional Neural Network Methods

The following demonstration illustrates using conventional neural network modeling methods to model data. The neural network will have seven hidden layers but will fail to sufficiently discriminate between the target events and nonevents in the data. The **develop** data are used in this demonstration. The data have been partitioned into three data sets (train, validation, and test) using a stratified random sample on the target variable and are described below.

A target marketing campaign for a bank was undertaken to identify segments of customers who are likely to respond to a variable annuity (an insurance product) marketing campaign. Each data set contains banking customers and 47 inputs that describe each customer. The 47 input variables represent other product usage in a three-month period and demographics. Two of the inputs are nominally scaled. The others are interval or binary. A binary target variable, **Ins**, indicates whether the customer bought the variable annuity product. The variables in the **develop** data set are listed below.

Table 1.4: Develop Data Set Variables

Variable	Description	Role	Level
ATM	Used ATM service (1=yes, 0=no)	Input	Binary
ATMAmt	ATM withdrawal amount	Input	Interval
AcctAge	Age of oldest account in years	Input	Interval
Age	Age of customer in years	Input	Interval
Branch	Branch of Bank (B1 – B19)	Rejected	Nominal
CC	Has credit card account (1=yes, 0=no)	Input	Binary
CCBal	Credit card balance	Input	Interval
CCPurc	Number of credit card purchases	Input	Interval
CD	Has certificate of deposit (1=yes, 0=no)	Input	Binary
CDBal	Certificate of deposit balance	Input	Interval
CRScore	Credit score	Input	Interval
CashBk	Number of times customer received cash back	Input	Interval
Checks	Number of checks	Input	Interval
DDA	Checking account (1=yes, 0=no)	Input	Binary
DDABal	Checking account balance	Input	Interval
Dep	Number of checking deposits	Input	Interval

Variable	Description	Role	Level
DepAmt	Amount deposited	Input	Interval
DirDep	Direct deposit (1=yes, 0=no)	Input	Binary
HMOwn	Owns home (1=yes, 0=no)	Input	Binary
HMVal	Home value in thousands of dollars	Input	Interval
ILS	Has installment loan (1=yes, 0=no)	Input	Binary
ILSBal	Installment loan balance	Input	Interval
IRA	Has retirement account (1=yes, 0=no)	Input	Binary
IRABal	Retirement account balance	Input	Interval
InArea	Local address (1=yes, 0=no)	Input	Binary
Income	Income in thousands of dollars	Input	Interval
Ins	Purchase variable annuity account (1=yes, 0=no)	Target	Binary
Inv	Has investment account (1=yes, 0=no)	Input	Binary
InvBal	Investment account balance	Input	Interval
LOC	Has line of credit (1=yes, 0=no)	Input	Binary
LOCBal	Line of credit balance	Input	Interval
LORes	Length of residence in years	Input	Interval
MM	Has money market account (1=yes, 0=no)	Input	Binary
MMBal	Money market balance	Input	Interval

Variable	Description	Role	Level
MMCred	Number of money market credits	Input	Interval
MTG	Has mortgage account (1=yes, 0=no)	Input	Binary
MTGBal	Mortgage balance	Input	Interval
Moved	Recent address change (1=yes, 0=no)	Input	Binary
NSF	Occurrence of insufficient funds (1=yes, 0=no)	Input	Binary
NSFAmt	Amount of insufficient funds	Input	Interval
POS	Number of point of sale transactions	Input	Interval
POSAmt	Amount in point of sale transactions	Input	Interval
Phone	Number of times customer used telephone banking	Input	Interval
Res	Area classification (R=rural, S=suburb, U=urban)	Rejected	Nominal
SDB	Has a safety deposit box (1=yes, 0=no)	Input	Binary
Sav	Saving account (1=yes, 0=no)	Input	Binary
SavBal	Saving balance	Input	Interval
Teller	Number of teller visits	Input	Interval

Here are the demonstration steps.

1. Open a browser window, navigate to SAS Studio, and sign in.
2. Expand **Server Files and Folders**.
3. Open the program titled **DLUS01D01a.sas**.
   The program opens in the code editor window. First, a caslib named mycas is created.
   ```
   libname mycas cas;
   ```
   Next, a local library named **local** is created.
   ```
   libname local '/home/student/LWDLUS';
   ```

Then, three DATA steps are used to create new data sets that are saved in memory.

```
data mycas.train_develop;
   set local.train_develop;
run;
data mycas.valid_develop;
   set local.valid_develop;
run;
data mycas.test_develop;
   set local.test_develop;
run;
```

4. Run the program.

5. Open the program titled **DLUS01D01b.sas**. First, the FREQ procedure is used to explore the binary outcome distribution in the validation partition of the **develop** data.

```
proc freq data=mycas.Valid_develop;
   table ins;
run;
```

Next, the NNET procedure is used to create a seven hidden-layer, feed-forward neural network. The *limited memory BFGS (L-BFGS)* is used to adjust the network's parameters. Like the original BFGS, L-BFGS uses an estimation of the inverse Hessian to steer the search. But whereas BFGS stores an *n-by-n* approximation to the Hessian (where *n* is the number of variables), the L-BFGS variant stores only a few vectors that represent the approximation implicitly.

```
proc nnet data=MYCAS.Train_DEVELOP standardize=std;
   target Ins / level=nominal;
   input AcctAge DDABal CashBk
         Checks NSFAmt Phone
         Teller SavBal ATMAmt
         POS POSAmt CDBal
         IRABal LOCBal ILSBal
         MMBal MMCred MTGBal
         CCBal CCPurc Income
      LORes HMVal Age
         CRScore Dep DepAmt InvBal / level=interval;
   input DDA DirDep NSF
         Sav ATM CD
         IRA LOC ILS
         MM MTG CC
         SDB HMOwn Moved
         InArea Inv / level=nominal;
   hidden 30;
   hidden 20;
   hidden 10;
   hidden 5;
   hidden 10;
   hidden 20;
   hidden 30;
   train outmodel=mycas._Nnet_model_
         validation=mycas.valid_develop numtries=1 seed=12345
         stagnation=15;
    optimization algorithm=LBFGS regL1=0.003 regL2=0.002
                 seed=12345 maxiter=50;
run;
```

6. Run the program and examine the results.

The results of PROC FREQ reveal that 34.63% of the customers responded to the marketing campaign.

Figure 1.10: Results of the FREQ Procedure

The FREQ Procedure

Ins	Frequency	Percent	Cumulative Frequency	Cumulative Percent
0	6327	65.37	6327	65.37
1	3352	34.63	9679	100.00

Next, the NNET procedure's results are shown. The results begin with a summary of the model information, followed by an Iteration History table.

Figure 1.11: Results of the NNET Procedure

The NNET Procedure

Model Information	
Model	Neural Net
Number of Observations Used	19358
Number of Observations Read	19358
Target/Response Variable	Ins
Number of Nodes	189
Number of Input Nodes	62
Number of Output Nodes	2
Number of Hidden Nodes	125
Number of Hidden Layers	7
Number of Weight Parameters	3590
Number of Bias Parameters	127
Architecture	MLP
Seed for Initial Weight	12345
Optimization Technique	LBFGS
Number of Neural Nets	1
Objective Value	2.772588103
Misclassification Rate for Validation	0.3463

						Norm			
Iteration Number	Objective Function	Norm of Gradient	Loss	Validate Error	Step Size	L1	L2	Maximum	Fit Error
1	2.967837	0.618776	2.772589	0.355822	0	64.00678	1.613834	0.199526	0.355719
2	2.848873	0.615200	2.772589	0.653683	5.522433	24.69071	1.105865	0.131593	0.653632
3	2.772588	0.944096	2.772588	0.346317	1	0	0	0	0.346368
4	2.721988	0.813089	2.705838	0.346317	0.015625	4.755949	0.940882	0.206158	0.346368
5	2.593922	0.020454	2.580768	0.346317	1	4.032717	0.527908	0.077512	0.346368
6	2.588562	0.062766	2.581512	0.346317	1	2.070418	0.419365	0.098395	0.346368
7	2.586168	0.020964	2.580770	0.346317	1	1.587692	0.317288	0.070472	0.346368
8	2.586119	0.019751	2.580761	0.346317	1	1.575310	0.316112	0.070472	0.346368
9	2.585443	0.079432	2.582049	0.346317	1	0.974456	0.235151	0.062727	0.346368
10	2.585445	0.079502	2.582052	0.346317	0.000488	0.974612	0.234369	0.062698	0.346368
11	2.585448	0.079594	2.582055	0.346317	0.000122	0.974799	0.234153	0.062619	0.346368
12	2.585450	0.079745	2.582060	0.346317	0.000061	0.974912	0.232621	0.062448	0.346368
13	2.585452	0.079891	2.582065	0.346317	7.629E-6	0.974904	0.230996	0.062262	0.346368
14	2.585454	0.080032	2.582070	0.346317	3.815E-6	0.974871	0.229627	0.062103	0.346368
15	2.585456	0.080143	2.582074	0.346317	3.815E-6	0.974965	0.228517	0.061971	0.346368
16	2.585458	0.080289	2.582079	0.346317	1.907E-6	0.975083	0.227058	0.061794	0.346368
17	2.585460	0.080407	2.582083	0.346317	9.537E-7	0.975178	0.225891	0.061649	0.346368
18	2.585462	0.080484	2.582086	0.346317	1.907E-6	0.975241	0.225130	0.061552	0.346368

Iteration History

It appears that the validation error (misclassification rate) in the last several iterations is equivalent to the empirical prior observed in the data! This shows that our model is not performing any better than a "coin toss" prediction.

The remaining results confirm the misclassification rate, the number of observations read, and the number of observations used by the neural network. A note in the output indicates that the model terminated the learning process on the 10th iteration due to lack of improvement on the validation assessment criteria.

Figure 1.12: Score Information

Score Information for Training	
Number of Observations Read	19358
Number of Observations Used	19358
Misclassification Rate	0.3464

Score Information for Validation	
Number of Observations Read	9679
Number of Observations Used	9679
Misclassification Rate	0.3463

The optimization exited on validation criteria.

Demonstration 2: Building and Training Deep Learning Neural Networks Using CASL Code

This demonstration illustrates building and training two deep learning neural networks. Both models will consist of the same number of hidden layers and hidden units as the neural network created in the last demonstration. However, deep learning techniques will be applied to these two new models.

This demonstration uses CASL code to implement deep learning models. PROC CAS is used to facilitate the use of CASL code.

1. Open the program titled **DLUS01D02a.sas** and examine the program's contents in the code editor window.

   First, the DeepLearn action set is loaded. This action set contains the bulk of SAS deep learning capabilities, including deep feed-forward (DNN), convolutional (CNN), and recurrent (RNN) neural networks.

   > **Note:** DeepLearn must be loaded before using SAS deep learning actions. The action set needs to be loaded only once.

   ```
   proc cas;
      loadactionset 'DeepLearn';
   run;
   ```

   Next, an empty deep learning model is created. The model's reference name is DLNN and the deep feed-forward model type is specified.

   ```
   proc cas;
      BuildModel / modeltable={name='DLNN', replace=1} type =
   "DNN";
      run;
   ```

Nine layers are then added to the model using the AddLayer action, the first of which is the input layer, as indicated by the TYPE='INPUT' option. Dropout is applied to the input layer using the DROPOUOT= option.

```
proc cas;
    AddLayer / model='DLNN' name='data' layer={type='input'
                STD='STD' dropout=.05};
```

Seven fully connected hidden layers are added. Each hidden layer is given a unique name in the NAME= option. The activation function used in each of these layers is the rectified linear (ReLU) activation function, with the exception of the first hidden layer. The first hidden layer incorporates the exponential linear (ELU) activation because ReLU is prone to saturation when connected with the input layer (communication with L. Lewis, SAS 2017). MSRA initialization is combined with the ReLU activation function and a dropout rate of 5% is included.

Dropout is applied to each of the hidden layers except for the middle hidden layer. Notice that the SRCLAYERS= option indicates the layer pertaining to the incoming connections.

```
AddLayer / model='DLNN' name='HLayer1'
layer={type='FULLCONNECT' n=30 act='ELU' init='xavier'
dropout=.05} srcLayers={'data'};
AddLayer / model='DLNN' name='HLayer2'
layer={type='FULLCONNECT' n=20 act='ReLU' init='MSRA'
dropout=.05} srcLayers={'HLayer1'};
AddLayer / model='DLNN' name='HLayer3'
layer={type='FULLCONNECT' n=10 act='ReLU' init='MSRA'
dropout=.05} srcLayers={'HLayer2'};
AddLayer / model='DLNN' name='HLayer4'
layer={type='FULLCONNECT' n=5 act='ReLU' init='MSRA'}
srcLayers={'HLayer3'};
AddLayer / model='DLNN' name='HLayer5'
layer={type='FULLCONNECT' n=10 act='ReLU' init='MSRA'
dropout=.05} srcLayers={'HLayer4'};
AddLayer / model='DLNN' name='HLayer6'
layer={type='FULLCONNECT' n=20 act='ReLU' init='MSRA'
dropout=.05} srcLayers={'HLayer5'};
AddLayer / model='DLNN' name='HLayer7'
layer={type='FULLCONNECT' n=30 act='ReLU' init='MSRA'
dropout=.05} srcLayers={'HLayer6'};
```

Last, an output layer is added. The softmax activation function is used because the target is binary.

```
AddLayer / model='DLNN' name="outlayer" layer={type='output'
          act='softmax'} srcLayers={"HLayer7"};
run;
```

Figure 1.13 is an abbreviated transcription of the model architecture constructed by the CASL code above.

Figure 1.13: Transcription of the Model Architecture

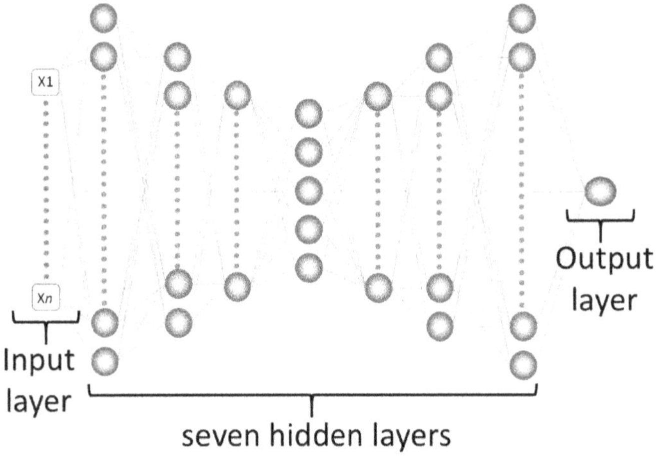

The model has now been constructed. However, we have not yet trained the model on our data. The dlTrain action is used to train the model.

The TABLE= option is populated with the name of the training data set loaded into CAS. The name of the model that we built is inserted between quotation marks in the MODEL= option.

The BESTWEIGHTS= option creates a data set that contains the weights corresponding to the model's best performance. If a validation data is included, then the best performing weights are chosen based on the model's performance on the validation data. Otherwise, the training data are used.

```
proc cas;
    dlTrain / table='Train_Develop' model='DLNN'
              bestweights={name='bestdeepweights', replace=1}
```

The list of inputs is specified in the INPUTS= option.

```
inputs={        'AcctAge',
                'DDABal',
                'CashBk',
                'Checks',
                'NSFAmt',
                'Phone',
                'Teller',
                'SavBal',
                'ATMAmt',
                'POS',
                'POSAmt',
                'CDBal',
                'IRABal',
                'LOCBal',
                'ILSBal',
                'MMBal',
                'MMCred',
                'MTGBal',
                'CCBal',
                'CCPurc',
                'Income',
```

```
                    'LORes',
                    'HMVal',
                    'Age',
                    'CRScore',
                    'Dep',
                    'DepAmt',
                    'InvBal',
                    'DDA',
                    'DirDep',
                    'NSF',
                    'Sav',
                    'ATM',
                    'CD',
                    'IRA',
                    'LOC',
                    'ILS',
                    'MM',
                    'MTG',
                    'CC',
                    'SDB',
                    'HMOwn',
                    'Moved',
                    'InArea',
                    'Inv'
                    }
```

The nominal inputs are specified in the NOMINAL= option.

```
        nominals={        'INS',
                  'DDA',
                    'DirDep',
                    'NSF',
                    'Sav',
                    'ATM',
                    'CD',
                    'IRA',
                    'LOC',
                    'ILS',
                    'MM',
                    'MTG',
                    'CC',
                    'SDB',
                    'HMOwn',
                    'Moved',
                    'InArea',
                  'Inv'
                    }
```

Note: Notice that the nominal variables must be specified in ***both*** the INPUT= and NOMINAL= options. Also, the target is binary and therefore must be specified in the nominals list.

The validation data table name is specified in the VALIDTABLE= option, and the INS target is included in the TARGET= option.

```
        ValidTable='Valid_Develop'
        target="INS"
```

Finally, the optimizer options are included. The ADAM method is used with a minibatch size of 60 and a learning rate of .001. The LRPOLICY= option specifies the learning rate policy—that is, strategy used to reduce the learning rate throughout the training process. The STEP policy multiplies the GAMMA= value by the current learning rate periodically, as indicated in the STEPSIZE= option. If no step size is indicated, then the learning rate is reduced every 10 epochs.

```
optimizer=    {minibatchsize=60,
         algorithm={method='ADAM',
                       lrpolicy='Step',
                       gamma=0.5,
                         stepsize=10,
                         beta1=0.9,beta2=0.999,
                         learningrate=.001}
         regL1=0.003,
            regL2=0.002,
            maxepochs=50
            }
seed=12345
;
run;
```

> **Note:** The MINIBATCHSIZE= option specifies the minibatch size <u>per worker</u>. The examples in this book were written to be delivered on a machine using 16 CPUs. Therefore, each total minibatch is actually the minibatch multiplied by the 16 CPUs. Sometimes long tails (fewer observations used in later iterations within an epoch) can form, which should be mitigated when possible. Include the LOGLEVEL=3 option of the optimization property to print detailed optimization information in the log.

2. Run the program and view the results.

The first set of tables contains descriptive information pertaining to the model shell and layers included in the model.

Figure 1.14: Model Shell and Layer Information

Results from deepLearn.buildModel

Output CAS Tables			
CAS Library	Name	Number of Rows	Number of Columns
CASUSER(student)	dlnn	1	5

Results from deepLearn.addLayer

Output CAS Tables			
CAS Library	Name	Number of Rows	Number of Columns
CASUSER(student)	dlnn	11	5

...

Results from deepLearn.addLayer

Output CAS Tables			
CAS Library	Name	Number of Rows	Number of Columns
CASUSER(student)	dlnn	88	5

Results from deepLearn.addLayer

Output CAS Tables			
CAS Library	Name	Number of Rows	Number of Columns
CASUSER(student)	dlnn	101	5

The next table contains the model information.

Figure 1.15: Model Information

Model dlnn Information Details	
Model Name	dlnn
Model Type	Deep Neural Network
Number of Layers	9
Number of Input Layers	1
Number of Output Layers	1
Number of Fully Connected Layers	7
Number of Weight Parameters	3620
Number of Bias Parameters	127
Total Number of Model Parameters	3747
Approximate Memory Cost for Training (MB)	3

The model contains 3,747 parameters and nine layers (seven hidden, one input and one output).

The Optimization History table shows the learning rate, training loss (error function value), validation loss, validation error (misclassification rate), and training error.

Figure 1.15: Optimization History Table

Optimization History of Deep Learning Model for TRAIN_DEVELOP							
Epoch	Learning Rate	Loss	Valid Loss	Valid Error	L1 Norm	L2 Norm	Fit Error
0	0.001	0.6604118656	0.631917	0.346317	1.844511	0.422419	0.35422
1	0.001	0.6281876242	0.605358	0.346317	1.577656	0.356574	0.347143
2	0.001	0.607519256	0.580518	0.346317	1.341028	0.301702	0.346368
3	0.001	0.5893408698	0.566654	0.286083	1.135575	0.258225	0.326842
4	0.001	0.5814298506	0.56125	0.278128	0.960872	0.218745	0.291301
5	0.001	0.5767605441	0.559204	0.270689	0.816366	0.187832	0.286187
6	0.001	0.5724564392	0.556409	0.276475	0.699637	0.162728	0.279833
7	0.001	0.5700404159	0.553769	0.271412	0.608331	0.141782	0.277921
8	0.001	0.5679287253	0.552814	0.271412	0.53486	0.124109	0.279264
9	0.001	0.5661265324	0.550567	0.269553	0.473884	0.109216	0.279626
10	0.0005	0.5857045705	0.550506	0.26666	0.432183	0.099407	0.279368
11	0.0005	0.5639419867	0.549184	0.267283	0.407416	0.093681	0.276578
12	0.0005	0.5639934868	0.548812	0.264284	0.385119	0.088338	0.274822
13	0.0005	0.5622428954	0.547997	0.263457	0.364619	0.083407	0.273634
14	0.0005	0.561439275	0.547097	0.264387	0.345111	0.078785	0.273944
15	0.0005	0.5627157391	0.547178	0.262424	0.327247	0.074583	0.275597
16	0.0005	0.5612882413	0.547896	0.26263	0.309477	0.070532	0.276217
17	0.0005	0.5638433015	0.547497	0.26294	0.293038	0.066874	0.276475
18	0.0005	0.5635103398	0.546745	0.26356	0.277522	0.063497	0.275028
19	0.0005	0.5620711933	0.546124	0.263044	0.263338	0.060413	0.274667
20	0.00025	0.5620258848	0.545926	0.261907	0.252857	0.058202	0.274822
21	0.00025	0.5630475659	0.546177	0.263044	0.246324	0.056832	0.278748
22	0.00025	0.5611724776	0.545678	0.263354	0.239998	0.055504	0.274254

...

43	0.000063	0.5624889646	0.544509	0.262424	0.174539	0.041776	0.27694
44	0.000063	0.5600745636	0.544153	0.262734	0.173585	0.041584	0.276527
45	0.000063	0.5601867709	0.544388	0.262424	0.172668	0.0414	0.275907
46	0.000063	0.5622321942	0.544421	0.262734	0.171666	0.041192	0.276785
47	0.000063	0.5624650323	0.544393	0.26263	0.170692	0.040986	0.277663
48	0.000063	0.5604373589	0.544388	0.262217	0.169726	0.040785	0.274718
49	0.000063	0.5617255545	0.544382	0.26294	0.168779	0.040589	0.275803
50	0.000031	0.5627086515	0.544405	0.26294	0.168033	0.040442	0.276217

The neural network using conventional methods failed to discriminate between the events and nonevents in the data. However, the incorporation of deep learning methods has drastically improved the model's discriminate ability. The misclassification rate has dropped from approximately 34.5% to approximately 26%. The model's best performance occurred in the 28th epoch. Perhaps the model can be further improved with batch normalization.

3. Open the program titled **DLUS01D02b.sas** and examine the program's contents in the code editor window.

The next model uses the exact same structure as the previous model, but with two exceptions. Batch normalization is used in all hidden layers (except for the first hidden layer) in place of dropout, and the ReLU hidden layer activation functions have been replaced with the hyperbolic tangent (tanh) activations. The weight initialization has also been changed from MSRA to Xavier.

> **Note:** In order for the batch normalization computations to conform to those described in Sergey Ioffe and Christian Szegedy's Batch Normalization research, the source layer should have settings of ACT=identity and INCLUDEBIAS=False. The activation function that would normally have been specified in the source layer should instead be specified on the batch normalization layer.

Batch normalization is not used on the first hidden layer because the input layer is already normalized using the z-score standardization method.

> **Note:** DNN-type architectures in SAS do not include batch normalization layers. However, a CNN-type architecture can implement batch normalization and will be used in place of
> a DNN.

```
proc cas;
   BuildModel / modeltable={name='BatchDLNN', replace=1}
            type = 'CNN';
/*          INPUT Layer             */
   AddLayer / model='BatchDLNN' name='data'
layer={type='input'
            STD='STD' dropout=.05};
/*          FIRST HIDDEN LAYER       */
   AddLayer / model='BatchDLNN' name='HLayer1'
            layer={type='FULLCONNECT' n=30 act='ELU'
init='xavier'
            } srcLayers={'data'};
```

Notice that each hidden layer after the first hidden layer now consists of two AddLayer statements. The first ADDLAYER statement in the second hidden layer specifies the weight initialization method and the number of hidden units. No bias is used in this layer and the activation function is set to identity. In the second ADDLAYER statement, the TYPE= option is set to BATCHNORM. Also, notice that the nonlinear transformation is applied in the second ADDLAYER statement.

```
/*          SECOND HIDDEN LAYER       */
AddLayer / model='BatchDLNN' name='HLayer2'
layer={type='FULLCONNECT' n=20 act='identity' init='xavier'
includeBias=False} srcLayers={'HLayer1'};
AddLayer / model='BatchDLNN' name='BatchLayer2'
layer={type='BATCHNORM' act='TANH'} srcLayers={'HLayer2'};
/*          THIRD HIDDEN LAYER        */
```

```
AddLayer / model='BatchDLNN' name='HLayer3'
layer={type='FULLCONNECT' n=10 act='identity' init='xavier'
includeBias=False } srcLayers={'BatchLayer2'};
AddLayer / model='BatchDLNN' name='BatchLayer3'
layer={type='BATCHNORM' act='TANH'} srcLayers={'HLayer3'};

/*          FOURTH HIDDEN LAYER        */
AddLayer / model='BatchDLNN' name='HLayer4'
layer={type='FULLCONNECT' n=5 act='identity' init='xavier'
includeBias=False } srcLayers={'BatchLayer3'};
AddLayer / model='BatchDLNN' name='BatchLayer4'
layer={type='BATCHNORM' act='TANH'} srcLayers={'HLayer4'};
/*          FIFTH HIDDEN LAYER         */
AddLayer / model='BatchDLNN' name='HLayer5'
layer={type='FULLCONNECT' n=10 act='identity' init='xavier'
includeBias=False } srcLayers={'BatchLayer4'};
AddLayer / model='BatchDLNN' name='BatchLayer5'
layer={type='BATCHNORM' act='TANH'} srcLayers={'HLayer5'};
/*          SIXTH HIDDEN LAYER         */
AddLayer / model='BatchDLNN' name='HLayer6'
layer={type='FULLCONNECT' n=20 act='identity' init='xavier'
includeBias=False} srcLayers={'BatchLayer5'};
AddLayer / model='BatchDLNN' name="BatchLayer6"
layer={type='BATCHNORM' act='TANH'} srcLayers={'HLayer6'};
/*          SEVENTH HIDDEN LAYER       */
AddLayer / model='BatchDLNN' name='HLayer7'
layer={type='FULLCONNECT' n=30 act='identity' init='xavier'
includeBias=False } srcLayers={'BatchLayer6'};
AddLayer / model='BatchDLNN' name="BatchLayer7"
layer={type='BATCHNORM' act='TANH'} srcLayers={'HLayer7'};

AddLayer / model='BatchDLNN' name='outlayer'
layer={type='output' act='LOGISTIC'}
srcLayers={'BatchLayer7'};
run;
```

The dlTrain code is the same as before, with one exception. The MODEL= option now directs the action to the new model structure created above, BatchDLNN.

```
proc cas;
dlTrain / table='Train_Develop' model='BatchDLNN'
…
optimizer=       {minibatchsize=60,
         algorithm=   {method='ADAM',
                        lrpolicy='Step',
                        gamma=0.5,
                        stepsize=10,
                  beta1=0.9,
                             beta2=0.999,
              learningrate=.001}
            regL1=0.003,
            regL2=0.002,
       maxepochs=50
            }
seed=54321
;
run;
```

4. Run the program and examine the results.
5. Scroll down in the results window to the Model Information Details table.

    The total number of model parameters has increased with batch normalization because we have added an additional operation with accompanying learnable parameters.

Figure 1.16: Model Information Details

Model batchdlnn Information Details	
Model Name	batchdlnn
Model Type	Convolutional Neural Network
Number of Layers	15
Number of Input Layers	1
Number of Output Layers	1
Number of Convolutional Layers	0
Number of Pooling Layers	0
Number of Fully Connected Layers	7
Number of Batch Normalization Layers	6
Number of Weight Parameters	3620
Number of Bias Parameters	222
Total Number of Model Parameters	3842
Approximate Memory Cost for Training (MB)	3

The Optimization History table shows that the memory cost paid off because the model has improved the validation misclassification rate, further reducing the rate down to approximately 25.9% (observed in the 15th pepoch).

Chapter 2: Convolutional Neural Networks

Introduction to Convoluted Neural Networks

A *convolutional neural network* (*CNN*) is a type of neural network that is viewed to be computationally and statistically efficient (Goodfellow, Bengio, and Courville 2017). CNNs have popularized the field of computer vision through many successes in image classification, object detection, and semantic segmentation. CNNs can also be used for natural language processing or other tasks where information is spatially correlated. A typical CNN consists of five types of layers: input, convolution, pooling, fully connected, and output. However, other types of layers are usually incorporated in more advanced structures, including transpose convolutional, region of interest pooling, segmentation, and many others. Each type of layer has its own specific properties and functionalities.

Computational efficiency is achieved through the use of convolution and pooling layers. Neurons in a fully connected layer attach a unique parameter to each incoming column of information. Conversely, convolution layers require fewer parameters than a fully connected layer because the parameters are shared across columns. Pooling layers themselves do not contain parameters, but instead they combine columns with an output summary. Adding convolution and pooling layers also enhances the model's ability to capture important points in input distributions, increasing the model's statistical efficiency. The details of convolution and pooling layers are discussed throughout this chapter.

Convolutional neural networks have popularized image classification and object detection. However, CNNs have also been applied to other areas such as natural language processing and forecasting. Many of the successes in self-driving cars, fault detection, chatbots, and other applications are in large part built on the foundation of convolutional neural networks.

Input Layers

Convolutional neural networks (CNNs) are a class of artificial neural networks. Like regular neural networks, a CNN begins with an input layer, consists of multiple layers, and contains a number of neurons, as seen in Figure 2.1.

Figure 2.1: Convolutional Neural Network

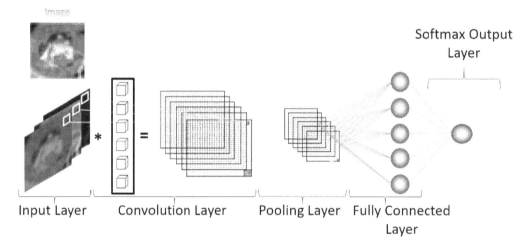

The input layer might consist of the raw pixel values of the image to be classified, while a grayscale image has only a single channel as seen in Figure 2.2.

Figure 2.2: Grayscale Image Channel

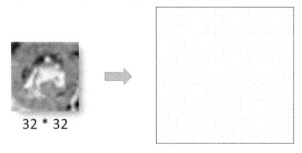

Conversely, a color image has three color channels of blue, green, and red as seen in Figure 2.3.

Figure 2.3: Color Image Channels

32 * 32 * 3

For example, each image in the CIFAR-10 data set has a height of 32, width of 32, and three color channels of blue, green, and red. Each color channel is represented by a matrix of values that indicate the brightness of the pixel with regard to the respective channel. Pixel values usually range from 0 to 255. Images are sometimes flattened before being passed to the convolutional neural network. However, flattening is not required and might even create problems when scoring with ASTORE. (ASTORE is a transportable form of a model called an analytic store.) Nonetheless, each pixel becomes a value in a respective column in a data set when an image is flattened. For example, a standard color image created by a standard 8k photo has more than 33 million pixels per channel. After it is flattened, the resulting table will consist of more than 99 million columns (~33 million * 3 channels).

CNNs in SAS can also contain input layers for the consumption of tabular data. Therefore, multiple input layers can be used. Quite often they are used when solving more dynamic problems.

Convolutional Layers

The convolution layer is most commonly used after the input layer. The convolution layer derives an output (that is, a feature map) from filter kernels that are connected to local regions in the input (that is, the incoming information). Each filter computes a dot product between their weights and a small region where they are connected to the input volume. The operation is formally known as a *cross-correlation* but might sometimes be referred to as a *convolution without the kernel flipping*. An example of single-channel convolution without kernel flipping is displayed below in Figure 2.4.

Figure 2.4: Single-channel Convolution Without Kernel Flipping

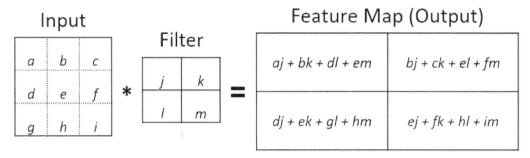

A typical convolution layer can consist of many filters. The filters slide across the input surface in parallel, capturing meaningful characteristics. Therefore, the parameters in each filter are shared by multiple columns in the data. Parameter sharing decreases the number of parameters needed to translate the input space. Each filter creates equivariant representations of the input, which means that changes in the input space are represented in the output.

Adding a convolution layer introduces new hyperparameters to the neural network. These hyperparameters include width, height, and stride. Here is an example of a convolution layer with 32 filters created in the SAS Cloud Analytic Services language (CASL):

```
AddLayer    /     model='ConVNet'
        name='ConVLayer1'
        layer={type='CONVO'
            nFilters=32
            width=3
            height=3
            stride=1
            }
        srcLayers={'data'};
```

Using Filters

It is quite common to increase the number of filters as the network becomes deeper. For example, the first convolutional layer might contain 32 filters, the next convolutional layer has 64 filters, the next has 128 filters, and so on. The number of convolutional filters is increased to offset the reduction of information that occurs when larger stride values are used (that is, stride values greater than one).

The following figures walk through the details of an example convolution layer.

During the forward pass of the convolution, the filter is placed in a starting position of the input space as seen in Figure 2.5.

Figure 2.5: Starting Position of the Filter

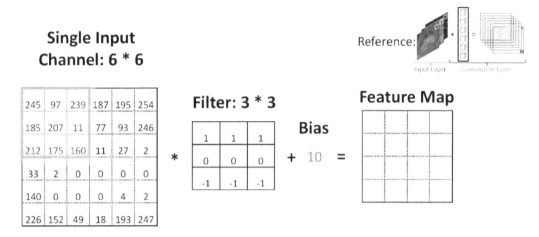

The dot products between the entries of the filter and the input at that position are calculated, as seen in Figure 2.6. For this example, we multiply 1*245 + 1*97 + 1*239 + 0*185 + 0*207 + 0*11 + -1*212 + -1*175 + -1*160 + our bias of 10, which gives us an output of 44 in the feature map.

Figure 2.6: Products of the Entries Between the Filter and Input

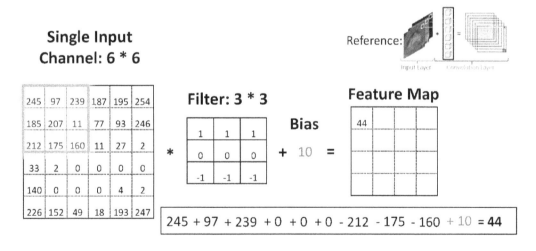

Therefore, the value derived from the dot product between the filter and the localized input space becomes the first output value. The algorithm then slides each filter across the input space. The range of movement is determined by the STRIDE hyperparameter, as seen in Figure 2.7. The stride value controls movement across the width of the input space, as well as the height.

Figure 2.7: Range Movement Due to STRIDE Hyperparameter

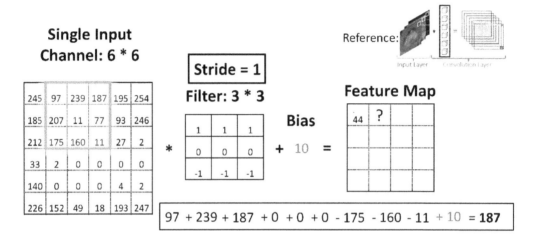

97 + 239 + 187 + 0 + 0 + 0 - 175 - 160 - 11 + 10 = **187**

After the filter has moved, the cross-correlation operation is continued, calculating the next output value. The process is repeated, and as the filter moves over the width and height of the input volume, we produce a two-dimensional feature map that gives the responses of that filter at every spatial position, as seen in Figure 2.8.

Figure 2.8: Feature Map with Filter Response at Every Spatial Position

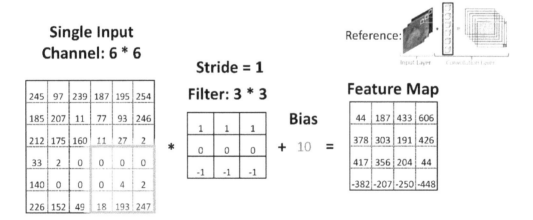

The filter values are the learnable parameters of the convolution layer (that is, the weights of the layer), with the bias equal to β.

Each filter will have the same number of channels as the incoming information. However, the output is always a two-dimensional grid, regardless of the number of channels. In the example presented in Figure 2.9, a single filter is combined with three channels of information. Each channel can be thought of as a "piece" of information. This single 3 by 3 filter has 28 weights. That is, each piece of information is assigned nine weights, which gives us 27 weights total. And then we have our bias, which gives us our 28th weight. A nonlinear activation $g(\cdot)$, such as an exponential linear, hyperbolic tangent, rectified linear, or other transformation, can be applied to the output values of the feature map, as seen in Figure 2.9.

Figure 2.9: Filter Weights and Nonlinear Transformation

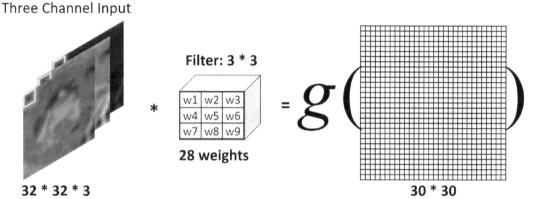

Three Channel Input

Filter: 3 * 3

w1 w2 w3
w4 w5 w6
w7 w8 w9

28 weights

32 * 32 * 3

30 * 30

Padding

Both convolution and pooling increase the pressures of underfitting due in large part to the reduction in information as more of these layers are added. Padding can be used to offset or mitigate the loss of information. Padding increases the relevance of pixels existing on the edge of an image.

Padding is calculated with the goal of producing an output image in a size that will be "reasonable" based on the original input image size and filter size. The key issues are whether the input image dimensions are even or odd, whether the filter dimensions are even or odd, and the value of the user-specified stride.

Here are some of the "reasonable" rules:

- The horizontal and vertical padding sizes are independent of each other. Changing the horizontal dimension of an image or filter has no effect on the vertical padding size.

- If the stride is 1, then the output image size is the same as the input image size. When discussing convolutions, this is sometimes referred to as *same padding*.

- If the stride is 2, then the output image size is about one half the input image size.

- Padding is first added to the right side of the image and then the left. So if padding is unequal, the right side will have more padding than the left.

- Images are padded with zeros.

Without padding, the feature map might be considerably smaller than the input, depending on the size of the filter and stride, as seen in Figure 2.10.

The shrinkage caused by convolution layers can be mitigated with the use of padding. If the stride is 1, the output map size is exactly the same as the input map size because the default behavior in the SAS DEEPLEARN action set is to automatically pad the channels or feature maps. If the stride is 2, the area of the output map is reduced by approximately 4 times (2*2).

Figure 2.10: Feature Map Without Padding

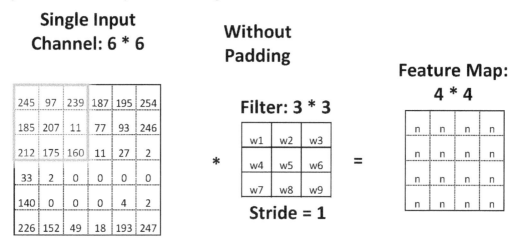

The shrinkage caused by convolution layers can be mitigated with the use of padding, as seen in Figure 2.11.

Figure 2.11: Feature Map with Padding

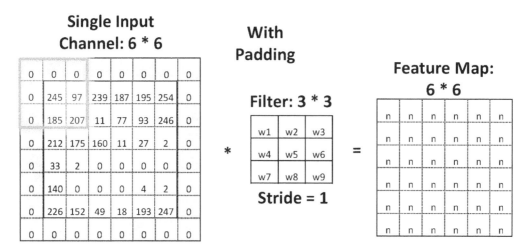

To use padding in SAS with GPUs, the FORCEEQUALPADDING option must be set to **True** in the dlTrain action. This means that equal padding is applied to all sides of the feature map. SAS defaults to FORCEEQUALPADDING=TRUE if the GPU=TRUE option is specified:

```
ForceEqualPadding=True
```

Without padding, the output feature map is reduced, as seen in Figure 2.12.

Figure 2.12: Without Padding

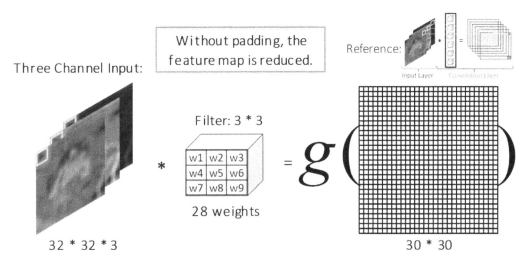

The dimension of the feature map is not what SAS produces by default, as shown next in Figure 2.13

Figure 2.13: Automatic Padding with SAS

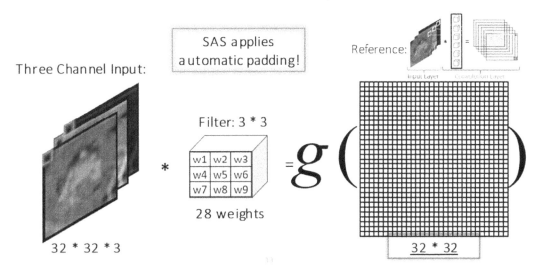

SAS automatically applies padding with the assumption the stride value equals one to offset the natural reduction in the feature map size. SAS still automatically applies padding when the stride value is greater than one, but the feature map will be smaller than the input matrix when the stride value is greater than one. Meaning using a stride value greater than one **will** decrease the output feature map size. Additionally, SAS automatically applies additional padding if the resulting convolution feature map is a noninteger matrix. For example, consider a scenario where the incoming information is a 12 x 12 matrix. Applying a filter of size 3 x 3 with a stride value of two yields an output of 6.5 (a noninteger value), as seen in Figure 2.14.

Figure 2.14: SAS Automatically Adjusts for Non-Integer Feature Maps

$$\left[\frac{12+2-3}{2} +1 \right] = 6.5 \quad \text{by} \quad \left[\frac{12+2-3}{2} +1 \right] = 6.5 . \quad \text{But SAS will:} \quad \left[\frac{12+3-3}{2} +1 \right] = 7$$

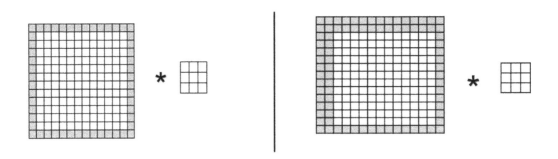

Feature Map Dimensions

The dimensions of the output are determined by the dimensions of the input, filter, and padding in combination with the stride size, as seen in Figure 2.15.

Figure 2.15: Feature Map Dimensions

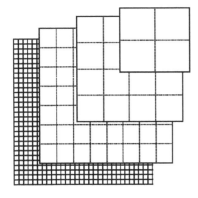

$$
\textbf{Output} \qquad\qquad \textbf{Output}
$$

$$
\left[\frac{i+2p-f}{s} +1 \right] \quad \textbf{X} \quad \left[\frac{i+2p-f}{s} +1 \right]
$$

where i **is the input dimension.**
p **is the padding dimension.**
f **is the filter dimension.**
s **is the stride.**

The feature map dimensions can be calculated by the following:

$$\textbf{Output} \qquad\qquad \textbf{Output}$$

$$\left[\frac{i + 2p - f}{s} + 1 \right] \quad \textbf{X} \quad \left[\frac{i + 2p - f}{s} + 1 \right]$$

i represents the input dimension, p represents the padding value (padding is discussed shortly), f represents the filter size, and S represents the stride size.

Pooling Layers

A pooling function provides a summary of some localized region of the incoming information. Three summary options are available in SAS: Maximum, Average, and Minimum. Similar to the convolution layer, the pooling layer uses filters, but the filters are used to establish the area to be summarized and do not contain learnable weights. The hyperparameters associated with the pooling filters include width, height, and stride. There is no need to specify the number of pooling filters because the filter count is determined by the number of incoming two-dimensional grids. It is common to periodically insert a pooling layer between successive convolution layers in a CNN architecture. Although, Springenberg et al. showed great success replacing pooling layers with convolution layers and leveraged stride a larger stride to down-sample the information (Springenberg et al., 2015). For example, a pooling or convolution layer with filters of size 2 x 2 applied with a stride of 2 reduces the amount of incoming information by 75%. Therefore, stride aggressively shrinks the network.

Generating a summary of each localized region has the added benefit of making the output approximately invariant to small deviations in the input. That is, the pooling function is useful when the modeler is more concerned with whether an object exists instead of its exact location. Conversely, if spatial differences are of great concern, then pooling should be used carefully, if even at all (Goodfellow, Bengio, and Courville 2017).

Pooling with larger stride values can also be used in situations where the size of your input information can vary. Consider a scenario where you are classifying images of two different sizes: 128 x 128 and 192 x 192. Each image size would be given its own input layer, and the larger image size can be down-sampled using pooling layers with stride greater than one to match the smaller input size. Down-sampling larger information is important because the output layer requires information to be of the same size.

The hyperparameters Width and Height set the size of the neighborhood, as seen in Figure 2.16.

Figure 2.16: Pooling Layers

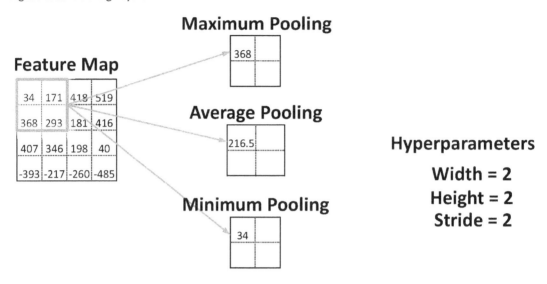

Each pooling function summarizes the neighborhood under examination. The POOL= option is used to specify the type of summary to be used by the pooling layer. Here is an example:

```
layer={type='POOL' width=2 height=2 stride=2 pool="max"
```

Just as with the convolution layers, the stride parameter controls the movement size of the neighborhood. The neighborhood continues to move across the input space until the entire space has been summarized, as seen in Figures 2.17 and 2.18.

Figure 2.17: Feature Map with Stride = 2

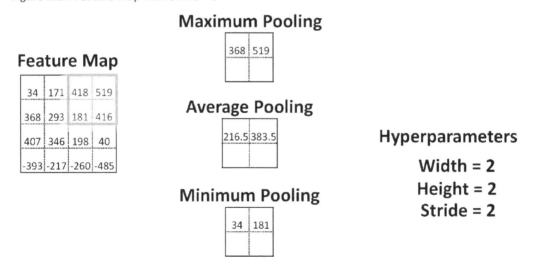

Figure 2.18: Completed Feature Map

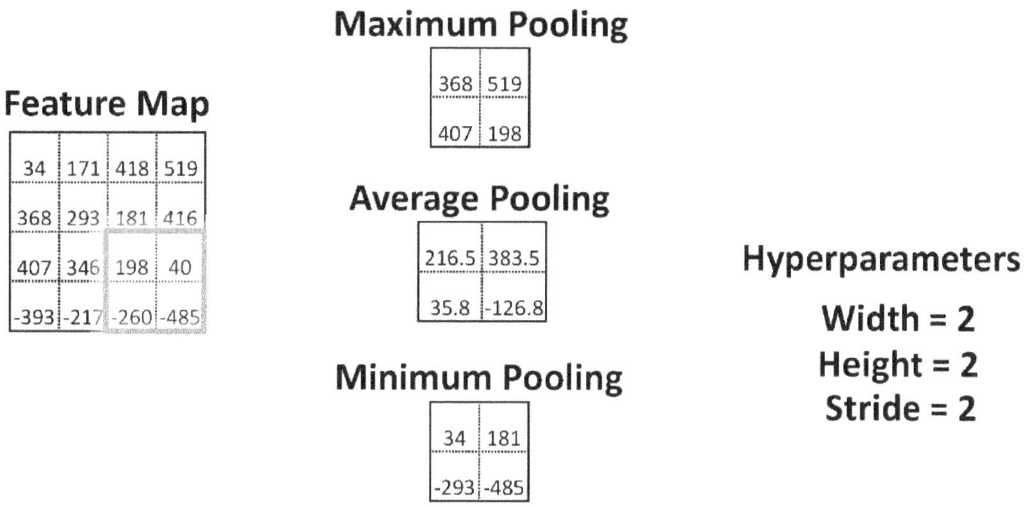

Maximum Pooling

368	519
407	198

Feature Map

34	171	418	519
368	293	181	416
407	346	198	40
-393	-217	-260	-485

Average Pooling

216.5	383.5
35.8	-126.8

Hyperparameters

Width = 2
Height = 2
Stride = 2

Minimum Pooling

34	181
-293	-485

Traditional Layers

Fully Connected Layer

Neurons in a fully connected layer have full connections to all activations in the previous layer, as seen in regular neural networks. Their activations can therefore be computed with a matrix multiplication followed by a bias offset. A fully connected layer incorporates a large number of parameters and therefore is expensive to train.

Output Layer

The output layer is essentially a fully connected layer that is associated with a particular error function. In this case of a binary target, the cross entropy error function

$$Q(\mathbf{w}) = 2\left[y \ln\left(\frac{y}{\mu(\mathbf{w})} \right) + (1-y) / \ln\left(\frac{1-y}{1-\mu(\mathbf{w})} \right) \right]$$

is simplified to an equivalent Bernoulli error function:

$$Q(\mathbf{w}) = -2\sum_{i}^{n} [\log(\hat{p}) + (1-y)\log(1-\hat{p})]$$

If the target's outcome has more than two levels, then the error function resolves to

$$\sum_{i}^{n}\sum_{c}^{C}-y_{true}^{(c)}\log(\hat{p}_{predicted}^{(c)})$$

where C is the class label for observation i.

If the target is binary, the logistic activation function is used:

$$\text{logistic}(net)=\frac{1}{1+e^{-net}}=\hat{p}\text{, where }net=w_{0}+\sum_{i=1}^{d}w_{i}x_{i}$$

The logistic activation function constrains its output to the range 0:1, making it ideal for generating probability ($\hat{p}$) estimates. In statistics, the logistic function is better known as the *logit-link* function:

$$\text{logit}(\hat{p})=\ln(\frac{\hat{p}}{1-\hat{p}})=\ln(odds)$$

If you are fitting a multinomial target, the *softmax* activation function (see below) is appropriate. It is the inverse of the *generalized logit-link* function.

$$\text{softmax}(net_{i})=\frac{e^{net_{i}}}{\sum_{j}e^{net_{j}}}$$

Because softmax divides the output activation of each neuron by the sum of the output by all participating neurons, it ensures that the estimates sum to 1. This produces a distributed effect.

Types of Skip-Layer Connections

Deep convolutional neural networks developed for image classification exploit the ability to enrich features by increasing the number of stacked layers. However, adding more layers might sometimes cause information learned early in the network to be forgotten as the gradient vanishes. Normalization techniques such as batch normalization have been successful in damping the wild gradient swings, largely addressing the convergence problems. Furthermore, adding skip-layer connections can help the neural network remember latent features learned early in the structure. A visualization of a skip-layer connection can be seen in Figure 2.19.

Figure 2.19: Skip-Layer Connection

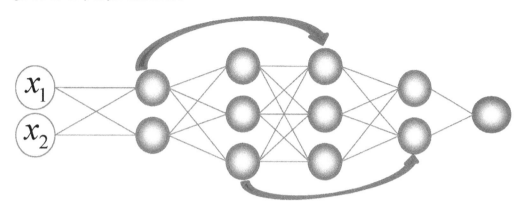

Concatenation and *residual* layers can be leveraged to create skip-layer connections. Concatenation layers can be used to create skip-layer connections (see Figure 2.20), but the concatenation operation results in "fat" networks, which are expensive to process.

```
AddLayer / model='ModelName' name='LayerName' layer={type='concat'}
          srcLayers={'A','B','C','D'};
```

Figure 2.20: Concatenation Layers

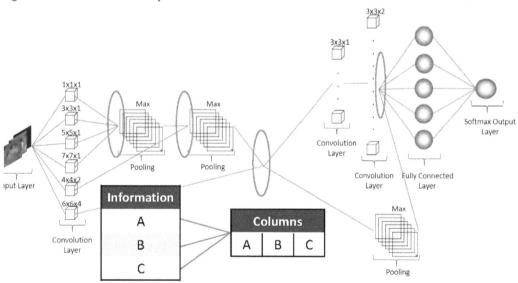

Residual layers were introduced by Kaiming He et al. as a means of combining information from skip-layer connections, as seen in Figure 2.21. Residual layers "thin out" the network because matching columns of information are summed, as opposed to concatenated. Use TYPE= 'RESIDUAL' to add a residual layer.

```
AddLayer / model='ModelName' name='LayerName'
          layer={type='residual'} srcLayers={'A','B','C','D'};
```

Figure 2.21: Residual Layers

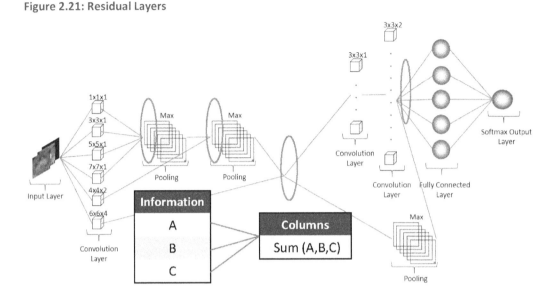

Demonstration 1: Loading and Preparing Image Data

This demonstration illustrates loading and preparing image data using the SAS Image action set.

1. Open the program titled **DLUS02D01.sas** and examine the program contents in the code editor window. First, the Image action set is loaded, and a metadata table describing the images is created. The loadImages action reads images and associated metadata into memory. The loadImages action can upload information from a variety of sources, including these:

    ○ photographic image files

    ○ biomedical image files, including those in Digital Imaging and Communications in Medicine (DICOM) format

    ○ ZIP files

    ○ a directory that contains image files or ZIP files

    ○ a directory that contains one or more series of photographic or biomedical images

    ○ a directory tree that contains any number of the preceding items

    ○ URLs

    ○ lists that contain combinations of the sources in the preceding items

The output contains images in a CAS table in encoded or decoded format, depending on the value of the **decode** input parameter. When this parameter is set to **False** (default value), the output table has the following columns:

○ **_id_** – an integer that is unique for each image

○ **_path_** – the full pathname of the image

○ **_image_** – a binary large object that contains the entire image file

○ **_size_** – the byte length of the binary large object in the **_image_** column

○ **_type_** – a three-character string that specifies the format of the image (for example, **jpg** for JPG images)

When the **decode** parameter is set to **True**, the **_image_** column contains decompressed image data, and the output table has these additional columns:

○ **_dimension_** – number of dimensions of the image (for example, 2 for two-dimensional images, 3 for three-dimensional images)

○ **_resolution_** – size of the image in each dimension (for example, {100, 200} for an array for a two-dimensional image that is 100 pixels wide and 200 pixels high)

○ **_imageFormat_** – integer that represents the organization of data in the **_image_** column. The most common values are 0, 16, 2, and 18, corresponding to 8-bit gray scale, 8-bit RGB, 16-bit gray scale, and 16-bit RGB images, respectively.

The LargetrainData and SmalltrainData folders located in the Image Data folder both contain 10 subfolders. Each of the 10 subfolders contains images of a specific class. For example, the truck folder contains images of only trucks. The RECURSE=TRUE option specifies the action to read in all the images from the 10 directories. The LABELLEVELS=1 option tells the action to assign the label based on the name of the folder where the image is stored. For example, all of the images in the truck folder will be assigned the label **truck**.

```
proc cas;
   loadactionset 'image';
   loadactionset 'table';
   table.addCaslib / name='imagelib'
      path='/home/student/LWDLUS/Image Data'
      subdirectories=true;

   image.loadimages / caslib='imagelib' path='LargetrainData'
         decode=true recurse=true labellevels=1
         addcolumns={'CHANNELCOUNT'}
            casout={name='LargetrainData', replace=true};

   image.loadimages / caslib='imagelib' path='SmalltrainData'
         decode=true recurse=true labellevels=1
         addcolumns={'CHANNELCOUNT'}
            casout={name='SmalltrainData', replace=true};
run;
quit;
```

Next, SAS Component Language is used to resize to a width and height of 120, and view images from several folders.

```
/***************/
/* View Images */
/***************/
data _NULL_;
/*AIRPLANES*/
  dcl odsout obj1();
  obj1.image(file:'/home/student/LWDLUS/Image
Data/SmalltrainData/airplane/img3.png',
                 width: "120",
                 height: "120");
  obj1.image(file:'/home/student/LWDLUS/Image
Data/SmalltrainData/airplane/img10.png',
                 width: "120",
                 height: "120");
  obj1.image(file:'/home/student/LWDLUS/Image
Data/SmalltrainData/airplane/img21.png',
                 width: "120",
                 height: "120");
 /*AUTOMOBILES*/
 dcl odsout obj2();
  obj2.image(file:'/home/student/LWDLUS/Image
Data/SmalltrainData/automobile/img6.png',
                 width: "120",
                 height: "120");
  obj2.image(file:'/home/student/LWDLUS/Image
Data/SmalltrainData/automobile/img9.png',
                 width: "120",
                 height: "120");
  obj2.image(file:'/home/student/LWDLUS/Image
Data/SmalltrainData/automobile/img37.png',
                 width: "120",
                 height: "120");
 /*BIRDS*/
  dcl odsout obj3();
  obj3.image(file:'/home/student/LWDLUS/Image
Data/SmalltrainData/bird/img25.png',
                 width: "120",
                 height: "120");
  obj3.image(file:'/home/student/LWDLUS/Image
Data/SmalltrainData/bird/img35.png',
                 width: "120",
                 height: "120");
  obj3.image(file:'/home/student/LWDLUS/Image
Data/SmalltrainData/bird/img65.png',
                 width: "120",
                 height: "120");
 /*CATS*/
  dcl odsout obj4();
  obj4.image(file:'/home/student/LWDLUS/Image
Data/SmalltrainData/cat/img0.png',
                 width: "120",
                 height: "120");
  obj4.image(file:'/home/student/LWDLUS/Image
Data/SmalltrainData/cat/img8.png',
                 width: "120",
                 height: "120");
```

```
    obj4.image(file:'/home/student/LWDLUS/Image
Data/SmalltrainData/cat/img46.png',
              width: "120",
              height: "120");

run;
```

Demonstration 2: Building and Training a Convolutional Neural Network

In this demonstration, a convolutional neural network is built and trained to classify images of 10 possible classes. A subset of the famous CIFAR-10 data is used. The data consist of 32 x 32 color images.

Open the program titled **DLMS02D02a.sas** and examine the program contents in the code editor window.

Examining the Image Data and Specifying the Model

1. First, the image data are summarized. The summary provides an average intensity for each color channel: blue, green, and red.

    ```
    proc cas;
        image.summarizeimages / table={name='LargeImageDatashuffled',
                                where='_PartInd_=1'};
    run;
    ```

Image column	PNG images count	Average intensity of B	Average intensity of G	Average intensity of R	Minimum intensity of R	Maximum intensity of R
image	40000	113.852228	123.021097	125.294747	0	255.000000

    ... .

    The summarized values are subtracted from the values in each input channel using the OFFSETS= option.

2. Second, the BuildModel action is specified, and an empty deep learning convolutional model is created.

    ```
    proc cas;
    BuildModel / modeltable={name='ConVNN', replace=1} type = 'CNN';
    ```

    a. The first layer added is an input layer. The number of channels is set to 3 in the NCHANNELS= option. The width and height of the image is also specified in the input layer. The offsets are also applied.

    ```
    AddLayer / model='ConVNN' name='data' layer={type='input'
      nchannels=3 width=32 height=32 offsets={113.852228,
      123.021097,125.294747}};
    ```

b. Next, a convolutional layer is added to the data. The convolutional layer contains six filters, each of a different size (1x1, 3x3, 5x5, 7x7, 4x4, and 6x6). The odd-value-sized filters use a stride of 1, the 4x4-sized filters use a stride of 2, and the 6x6-sized filters use a stride of 4. The 4x4 and 6x6 sets include two extra filters. In addition, dropout is applied to both the 4x4 and 6x6 sets of filters.

```
AddLayer / model='ConVNN' name='ConVLayer1a'
layer={type='CONVO'
          nFilters=8  width=1 height=1 stride=1}
          srcLayers={'data'};

AddLayer / model='ConVNN' name='ConVLayer1b'
layer={type='CONVO'
          nFilters=8  width=3 height=3 stride=1}
          srcLayers={'data'};

AddLayer / model='ConVNN' name='ConVLayer1c'
layer={type='CONVO'
          nFilters=8  width=5 height=5 stride=1}
          srcLayers={'data'};

AddLayer / model='ConVNN' name='ConVLayer1d'
layer={type='CONVO'
          nFilters=8  width=7 height=7 stride=1}
          srcLayers={'data'};

AddLayer / model='ConVNN' name='ConVLayer1e'
layer={type='CONVO'
          nFilters=10  width=4 height=4 stride=2 dropout=.2}
          srcLayers={'data'};

AddLayer / model='ConVNN' name='ConVLayer1f'
layer={type='CONVO'
          nFilters=10  width=6 height=6 stride=4 dropout=.2}
          srcLayers={'data'};
```

c. A concatenation layer is used to combine the four paths into a single path in preparation for the next layer. Specifically, the convolutions that use a stride of 1 are combined and connected to the next layer. The other convolutions will be connected to later layers.

```
AddLayer / model='ConVNN' name='concatlayer1a'
          layer={type='concat'}

srcLayers={'ConVLayer1a','ConVLayer1b','ConVLayer1c',
                    'ConVLayer1d'};
```

d. The concatenation layer is connected to a pooling layer that is set to extract the maximum value from each neighborhood. The pooling layer creates a 2 x 2 neighborhood that moves across columns with a stride of 2.

```
AddLayer / model='ConVNN' name='PoolLayer1max'
layer={type='POOL'
          width=2 height=2 stride=2 pool='max'}
          srcLayers={'concatlayer1a'};
```

e. A concatenation layer is used to combine the pooling layer with the convolutional layer that is connected to the input layer and uses a 4 x 4 filter.

> **Note:** Concatenation layers require that the incoming feature maps be the same size. It is a common practice to use pooling layers to down-sample larger feature maps when combining feature maps of varying sizes.

```
AddLayer / model='ConVNN' name='concatlayer2'
            layer={type='concat'}
            srcLayers={'PoolLayer1max','ConVLayer1e'};
```

f. The concatenation layer is then connected to another 2 x 2 pooling layer with a stride of 2, max pooling layer.

```
AddLayer / model='ConVNN' name='PoolLayer2max'
layer={type='POOL'
            width=2 height=2 stride=2 pool='max'}
            srcLayers={'concatlayer2'};
```

g. A concatenation layer is used to combine the pooling layer with the convolutional layer that is connected to the input layer and uses a 6 x 6 filter.

```
AddLayer / model='ConVNN' name='concatlayer3'
            layer={type='concat'}
            srcLayers={'PoolLayer2max','ConVLayer1f'};
```

h. The concatenation layer is then connected to another 2 x 2 pooling layer with a stride of 2, max pooling layer.

```
AddLayer / model='ConVNN' name='PoolLayer3max'
layer={type='POOL'
            width=2 height=2 stride=2 pool='max'}
            srcLayers={'concatlayer3'};
```

i. The pooling layer is then connected to a convolution layer with 64 filters. The convolution layer includes 3 x 3 filters that move across the input columns using a stride of 1. The MSRA2 is used to initialize the weights. A dropout rate of 20% is also applied to the convolution layer. Typically, the hidden bias would be removed, and the activation function would be set to **Identity** before applying batch normalization. However, in an attempt to show the flexibility of SAS, we have decided to not follow conventional methods.

```
AddLayer / model='ConVNN' name='ConVLayer1g'
layer={type='CONVO'
            nFilters=64 width=3 height=3 stride=1 init='msra2'
            dropout=.2} srcLayers={'concatlayer3'};

AddLayer / model='ConVNN' name='BatchLayer1'
            layer={type='BATCHNORM' act='ELU'}
            srcLayers={'ConVLayer1g'};
```

j. The normalized values are then passed to another convolution layer with 128 filters. The convolution layer includes 3 x 3 filters that move across the input columns using a stride of 2. MSRA2, batch normalization, and dropout are also used in this convolution layer.

```
AddLayer / model='ConVNN' name='ConVLayer1h'
layer={type='CONVO'
            nFilters=128 width=3 height=3 stride=2 init='msra2'
            dropout=.2} srcLayers={'BatchLayer1'};
```

```
AddLayer / model='ConVNN' name='BatchLayer2'
          layer={type='BATCHNORM' act='ELU'}
          srcLayers={'ConVLayer1h'};
```

k. A concatenation layer is then used to concatenate the convolutional layer with the last applied pooling layer.

```
AddLayer / model='ConVNN' name='concatlayer4'
          layer={type='concat'}
          srcLayers={'PoolLayer3max','BatchLayer2'};
```

l. A fully connected layer is then added with 240 neurons. Batch normalization and a dropout rate of 65% are applied to the fully connected layer.

```
AddLayer / model='ConVNN' name='FCLayer2'
          layer={type='FULLCONNECT' n=240 act='Identity'
          init='msra2' dropout=.65 includeBias=False}
          srcLayers={'concatlayer4'};

AddLayer / model='ConVNN' name='BatchLayer3'
          layer={type='BATCHNORM' act='ELU'}
          srcLayers={'FCLayer2'};
```

m. Finally, the normalized values are passed to the output layer from the previous fully connected layer.

```
AddLayer / model='ConVNN' name='outlayer' layer={type='output'
          act='SOFTMAX'} srcLayers={'BatchLayer3'};
run;
```

> **Note:** The model generated by the code above resembles Figure 2.22 below.

Figure 2.22: Model Generated by Code

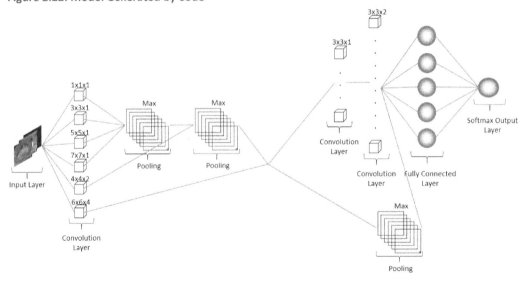

Fitting and Assessing the Model

1. The dlTrain action is called to train the model. The best weights are stored in a data set named **ConVbestweights**. The model will also be trained using a GPU instead of a CPU because of the large number of weights in the specified model. In addition, the table containing the model's performance for each epoch is saved in a table named **ObjectModeliter**. This table is used to create an iteration plot.

```
ods output OptIterHistory=ObjectModeliter;
proc cas;
    dlTrain / table={name='LargeImageDatashuffled',
        where='_PartInd_=1'} model='ConVNN'
        modelWeights={name='ConVTrainedWeights_d', replace=1}
        bestweights={name='ConVbestweights', replace=1}
        inputs='_image_'
        target='_label_' nominal={'_label_'}
        GPU=True
         ValidTable={name='LargeImageDatashuffled',
                    where='_PartInd_=2'}

    optimizer={minibatchsize=60,algorithm={method='ADAM',
    lrpolicy='Step', gamma=0.6, stepsize=5, beta1=0.9, beta2=0.999,
    learningrate=.01}maxepochs=60}
    seed=12345
;
run;
```

The remaining code creates two macro variables that store the lowest misclassification values for the training and validation data, respectively. PROC SGPLOT is called to plot out the iteration history for the validation and training misclassification rates, and macro variables containing the lowest misclassification values are used to label each curve.

```
proc sql noprint;
select min(FitError)
   into :Train separated by ' '
   from ObjectModeliter;
quit;

proc sql noprint;
select min(ValidError)
   into :Valid separated by ' '
   from ObjectModeliter;
quit;

proc sgplot data=ObjectModeliter;
   yaxis label='Misclassification Rate';
   series x=Epoch y=FitError / CURVELABEL="&Train"
          CURVELABELPOS=END;
   series x=Epoch y=ValidError / CURVELABEL="&Valid"
          CURVELABELPOS=END;
run;
```

2. View the results after the program finishes running.

   Scroll down in the Results window to the Model convnn Information Details table (Figure 2.23). Notice that the model contains 801,974 parameters!

   Figure 2.23: Model Convnn Information Details Table

Model convnn Information Details	
Model Name	convnn
Model Type	Convolutional Neural Network
Number of Layers	21
Number of Input Layers	1
Number of Output Layers	1
Number of Convolutional Layers	8
Number of Pooling Layers	3
Number of Fully Connected Layers	1
Number of Batch Normalization Layers	3
Number of Concatenation Layers	4
Number of Weight Parameters	800856
Number of Bias Parameters	1118
Total Number of Model Parameters	801974
Approximate Memory Cost for Training (MB)	28

   Next, scroll down to the Optimization History of Deep Learning Model for LARGEIMAGEDATASHUFFLED table. The best performance on the validation occurred at epoch 52. The model has a 23.73% misclassification rate on the validation data. Notice that there is considerable divergence between the training and validation misclassification rates. Perhaps a greater use of regularizations could push the validation error rate down further. This could include increasing the dropout rate, applying dropout to other layers, incorporating L1 or L2 regularizations, or generating synthetic cases with carefully chosen random mutations, as seen in Figure 2.24.

Figure 2.24: Misclassification Rate

Optimization History of Deep Learning Model for LARGEIMAGEDATASHUFFLED

Epoch	Learning Rate	Loss	Valid Loss	Valid Error	Fit Error
0	0.01	1.6941935995	1.402493	0.5134	0.619425
1	0.01	1.3818233751	1.414891	0.5101	0.5061
2	0.01	1.2584495667	1.366709	0.4847	0.454675
3	0.01	1.1553676277	1.072168	0.3819	0.415225
4	0.01	1.0853198762	1.027941	0.3648	0.3882
5	0.006	0.9775354026	0.981807	0.3467	0.3481

....

Epoch	Learning Rate	Loss	Valid Loss	Valid Error	Fit Error
51	0.00006	0.4552174167	0.767141	0.2377	0.162925
52	0.00006	0.4545912464	0.766216	0.2373	0.160525
53	0.00006	0.4523058415	0.76678	0.2377	0.160375
54	0.00006	0.4529939283	0.766223	0.2375	0.1603
55	0.000036	0.4541864682	0.766705	0.2374	0.16115
56	0.000036	0.4540002291	0.766814	0.2374	0.1602
57	0.000036	0.4492199562	0.76731	0.2377	0.15935
58	0.000036	0.4522716056	0.765956	0.2381	0.16145
59	0.000036	0.4534152824	0.767631	0.2382	0.16055

Note: Results might vary due to the distribution of data and computation.

The iteration plot in Figure 2.25 shows the performance of the model's misclassification rates on the training and validation data sets.

Figure 2.25: Performance of Model's Misclassification Rates

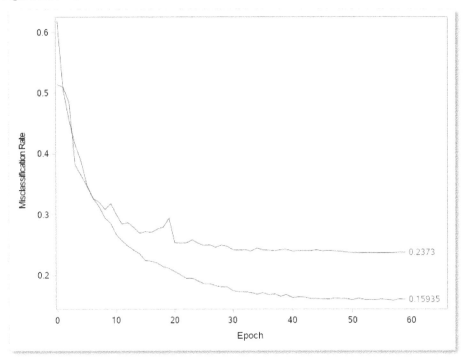

Scoring with the Fitted Model

1. Open the program named **DLMS02D02b.sas.** Now that we have our model, it is time to score new data. The dlScore action is used to score the validation data. Notice that the INITWEIGHTS= option specifies the best weights discovered by the trained model (that is, the weights discovered in epoch 55).

```
proc cas;
    dlScore / table={name='LargeImageDatashuffled',
                     where='_PartInd_=2'} model='ConVNet'
              initWeights='ConVbestweights'
              layerOut={name='Layer_data', replace=1}
              layers='ConVLayer1'
              layerImageType='JPG'
              casout={name='ScoredData', replace=1}
              copyVars='_Label_'
    ;
run;
proc print data=mycas.ScoredData (obs=20);
run;
```

> **Note:** dlScore applies SAS Enterprise Miner naming conventions to the prediction variables when the option ENCODENAME=TRUE is specified.

SAS detects the presence of the target variable and provides error statistics in the Score Information table (as seen in Figure 2.26).

Figure 2.26: Score Information Table

Score Information for SMALLIMAGEDATASHUFFLED	
Number of Observations Read	8000
Number of Observations Used	8000
Misclassification Error (%)	23.575
Loss Error	0.736012

The model appears to have a misclassification rate of 23.58 percent on the holdout data.

2. Run the remaining code in the program and view the results.

The program prints the first 20 observations from the score data created by the dlScore action. The **_label_** column contains the actual label, and the **I__label_** column contains the predicted classification. PROC SGPLOT is used to create a histogram showing the number of misclassified images for each class. (See Figure 2.27.)

Figure 2.27: Histogram of Misclassified Images

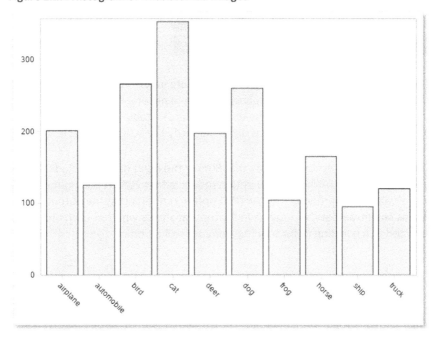

An example of the output produced by the code is displayed in Figure 2.28.

Figure 2.28: Code Output

The PARTITION procedure is used to sample data in SAS Viya. It performs simple random sampling, stratified sampling, oversampling, or *k*-fold partitioning to produce a table that contains a subset of the observations or a table that contains partitioned observations. The BY statement is used to specify the stratum variable. In this example, the target is used as the stratum.

The SAMPPCT=80 option requests that 80% of the input data be included in the training partition, and the SAMPPCT2=20 option requests that 20% of the input data be included in the validation partition. The PARTIND option requests that the output data table, **mycas.smalltraindata**, include an indicator that shows whether each observation is selected to a partition (1 for training and 2 for validation).

A second run of the PARTITION procedure is used to partition the larger sample of images in the same manner as described above.

```
proc partition data=mycas.SmalltrainData samppct=80
     samppct2=20 seed=12345 partind;
     by _label_;
   output out=mycas.smallImageData;
run;

proc partition data=mycas.LargetrainData samppct=80
       samppct2=20 seed=12345 partind;
     by _label_;
     output out=mycas.LargeImageData;
run;
```

Small Sample can be seen in Figure 2.29.

Figure 2.29: Small Sample

		Stratified Sampling Frequency		
Index	_label_	Number of Obs	Sample Size 1	Sample Size 2
0	airplane	1000	800	200
1	automobile	1000	800	200
2	bird	1000	800	200
3	cat	1000	800	200
4	deer	1000	800	200
5	dog	1000	800	200
6	frog	1000	800	200
7	horse	1000	800	200
8	ship	1000	800	200
9	truck	1000	800	200

Large Sample can be seen in Figure 2.30.

Figure 2.30: Large Sample

	Stratified Sampling Frequency			
Index	_label_	Number of Obs	Sample Size 1	Sample Size 2
0	airplane	5000	4000	1000
1	automobile	5000	4000	1000
2	bird	5000	4000	1000
3	cat	5000	4000	1000
4	deer	5000	4000	1000
5	dog	5000	4000	1000
6	frog	5000	4000	1000
7	horse	5000	4000	1000
8	ship	5000	4000	1000
9	truck	5000	4000	1000

The small sample can be used for practice if you want to run this example more quickly instead of using the large sample.

The data are currently ordered by the target values. For example, the data might begin with pictures of only frogs, followed by pictures of cats, and so on. Ordering by the target can cause a problem for variants of stochastic gradient descent (SGD, momentum, and ADAM). The problem arises because ordering by the outcome can cause the model to become entrenched with a set of parameters that overpredicts a single class. That is, it fails to discriminate between all outcome classes. It is recommended that the data be randomly shuffled. If fact, it is quite common to change the observation order when blending (ensemble through weighted averages of predictions) models together for image classifiers.

Note: An alternative to random shuffling is *curriculum learning*. Curriculum learning sorts the data by those observations that are easiest to learn. To implement curriculum learning, a practitioner would train the model, sort the data based on the predictions versus actuals, and then retrain the model.

The following code randomly sorts the observations using the shuffle action. A new data set named **ImageDataShuffled** is created.

```
proc cas;
 table.shuffle / table='smallImageData'
            casout={name='SmallImageDatashuffled', replace=1};
run;

proc cas;
 table.shuffle / table='LargeImageData'
            casout={name='LargeImageDatashuffled', replace=1};
run;
```

Convolutional neural networks can take days, weeks, or even months to train in certain situations. This example builds a simple convolutional network with a limited amount of data to save time.

3. Open the program titled **DLMS02D02a.sas** and run the program.

Chapter 3: Improving Accuracy

Introduction

Most traditional data such as transaction or survey data does not contain the "truth." For example, knowing the actual quality of an individual's moral fiber could dramatically improve the accuracy of a fraud model, although there are types of data in which the truth is contained. Image data is one example where the truth is present because the elements that define an object exist in the image, albeit two-dimensionally. It is therefore not surprising that a properly trained neural network can outperform a layperson in image classification and other vision-related tasks.

The focus of this chapter is to highlight simple techniques that can be deployed to improve computer vision models.

Architectural Design Strategies

A 1 x 1 convolution is commonly used to change the dimensional depth of information flowing through the network. A 1 x 1 kernel shares a single weight across all columns of all channels of information, which produces a summarized motif. (See Figure 3.1.)

Figure 3.1: 1 x 1 Convolutions

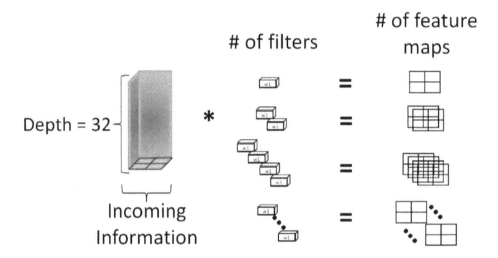

GoogLeNet deployed 1 x 1 convolutions to reduce the dimensional depth before applying larger kernel filters, a technique referred to as bottlenecking (Szegedy et al. 2014). Bottlenecking can reduce the computational cost associated with training, enabling resources to be allocated toward other areas, such as increasing cardinality or constructing a deeper, wider network structure. (See Figure 3.2.)

Figure 3.2: Dimensional Depth Reduction Before Feature Extraction

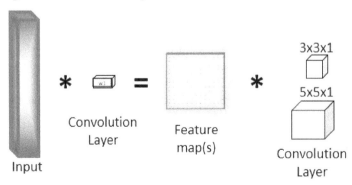

You can also use 1 x 1 convolutions to expand the dimensional depth, and they are sometimes used in place of fully connected layers to save on computational cost, as shown in Figure 3.3.

Figure 3.3: Using 1 x 1 Convolutions to Expand Dimensional Depth

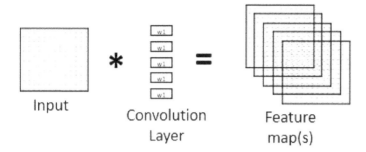

Spatial Exploration

Filters extract spatially correlated information and transcribe that information to an output feature map. Spatial exploration techniques such as the use of filters of different sizes enables the model to capture varying levels of granularity, providing a more complete understanding of the region for which the filters are located. The information captured at varying magnitudes can then be combined in subsequent layers to form a more detailed explanation of the information.

In addition, increasing the stride influences the exploration scheme of each filter and reduces the size of the output feature map. The width and height of output feature maps are reduced by larger stride values.

Creating Blocks

GoogLeNet combined 1 x 1 filters with spatial exploration to create *blocks* (Szegedy et al. 2014). A block is a stage of feature extraction that commonly consists of multiple convolutions that are usually stacked and then combined with either a concatenation layer or residual layer in a *split-transform-merge* strategy. (See Figure 3.4.)

Figure 3.4: A Block

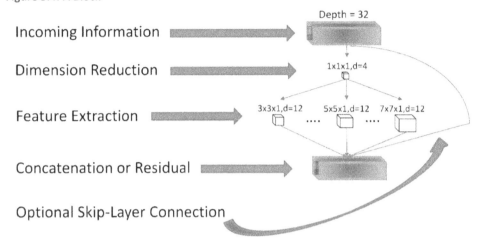

A block differs from a hidden layer because blocks can consist of multiple hidden layers that, when combined, produce a more relevant output. Sometimes pooling layers are used within a block, which was the situation in GoogLeNet. Blocks are usually stacked much like hidden layers.

ResNet introduced residual layers as a means of combining previously learned information with the current feature state to formulate a model that can effectively support increased depth (He et al. 2015). ResNet leveraged residual layers within what are called *residual blocks*.

Figure 3.5: ResNet Type Residual Block

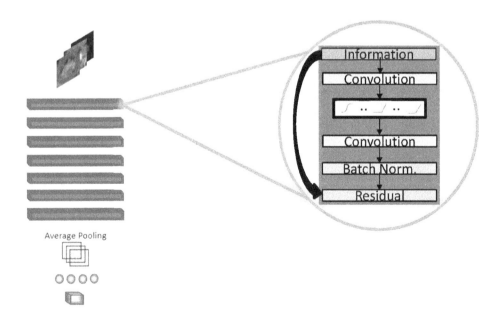

A residual block begins with a convolution layer applied to the incoming information. The output of the first convolution passes through a nonlinear activation before the next convolution layer. Batch normalization is used to prevent the gradient from exploding and is applied to normalize the information before another nonlinear activation is involved. The residual block closes with a skip-layer residual connection that combines the original information with the newly learned features. Residual blocks mitigate the vanishing/exploding gradient problem associated with increased network depth.

Cardinality is in reference to the number of transformation sets within a block. *Cardinality* was first introduced in the ResNeXt model in 2017 as an alternative to depth exploration for more accuracy (Saining et al. 2017). It can be expensive to train, but if skip layers are used, residual connections are recommended to reduce the training cost.

Figure 3.6: Cardinality

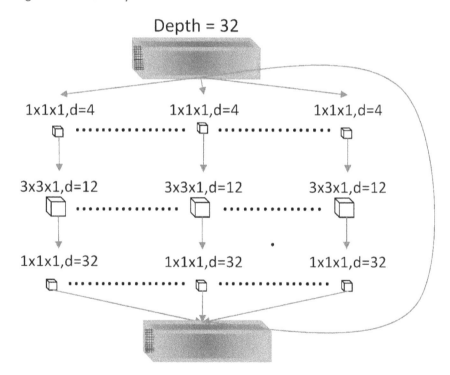

Each transformation set consists of an independent track of information that is trained jointly, not independently. For example, Figure 3.7 below displays five transformation sets within the block.

Figure 3.7: Transformation Sets Within a Block

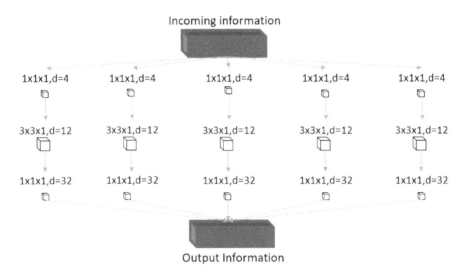

ResNeXt showed that cardinality can be used as an alternative to depth exploration. Specifically, the authors of ResNeXt demonstrated a lower misclassification rate compared to a ResNet with a similar number of parameters. Each transformation set used in ResNeXt begins with a 1 x 1 convolution to reduce the dimensional depth, followed with a 3 x 3 convolution used for feature extraction. Lastly, a 1 x 1 is used to expand the dimensional depth before the residual layer summarizes and combines all transformation sets.

Figure 3.8: ResNet Type with Average Pooling

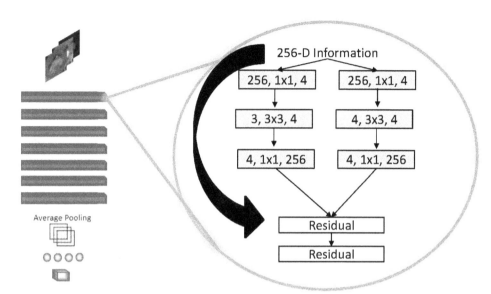

Interestingly, this author has observed cardinality to behave much like a structural regularity. This means that cardinality would be a useful addition to a model that seems to overfit the training data.

Comparing Structural Depth to Cardinality

Consider the following case study that examines the performance of a ResNet type model and ResNeXt type model on the CIFAR-10 data. Specifically, the impact of cardinality versus network depth is examined. Table 3.1 highlights some of the differences between ResNet-type and ResNext-type models.

Table 3.1: ResNet versus ResNext Architectures

Attribute	ResNet Type	ResNext Type
Number of Blocks	10	3
Cardinality	1	6
Skip-Layer Connections	residual	residual
Number of Parameters	803,230	349,842

Both the ResNet type and ResNeXt type models examined in this experiment have almost identical beginning and ending model structures. ResNet begins with a convolution layer containing *12* filters, followed by a 2 x 2 max pooling layer, and ends with a 3 x 3 average pooling layer followed by a fully connected layer with 660 neurons. ResNeXt begins with a convolution layer containing *8* filters, followed by a 2 x 2 max pooling layer, and ends with a 3 x 3 average pooling layer followed by a fully connected layer with 660 neurons.

The primary difference between the two architectures is the number of blocks and cardinality used between the beginning and ending layers. The ResNet type model uses 10 blocks with a single independent path, meaning that the cardinality is 1. The ResNeXt type is a much shallower model, using only three blocks with six separate tracks, meaning that the cardinality is 6.

Cardinality has an additional unforeseen computational cost. Therefore, despite the difference in model parameters, the ResNeXt type model takes slightly longer to train with fewer parameters than the ResNet type model.

First, both models are trained on 10,000 observations for 60 epochs. ResNet performs significantly better, with a misclassification rate of 44.2% compared to ResNeXt's misclassification rate of 48.55%. Interestingly, ResNeXt seems to have lower variance compared to ResNet. See Figure 3.9 and Table 3.2.

Figure 3.9: ResNet versus ResNext, 10,000 observations for 60 Epochs

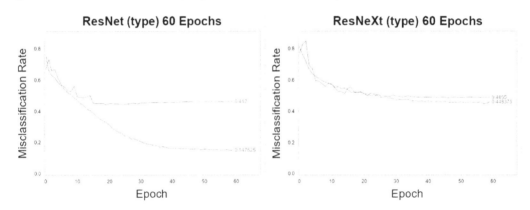

Table 3.2: ResNet versus ResNext, 10,000 Observations for 60 Epochs

Validation Misc	ResNet	ResNext
10k obs, 60 epochs	44.2	48.55

Epochs were then increased from 60 to 1500, and both models were retrained. Performance for both models improved, and the discrepancy between the two models is reduced, although the ResNet type model still outperformed the ResNeXt type model. Training for 1500 epochs seems a bit extreme given that the data contains only 10,000 observations. See Figure 3.10 and Table 3.3.

Figure 3.10: ResNet versus ResNext, 10,000 Observations for 1,500 Epochs

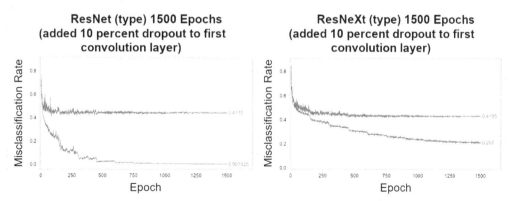

Table 3.3: ResNet versus ResNext, 10,000 Observations for 1,500 Epochs

Validation Misc	ResNet	ResNext
10k obs, 60 epochs	44.2	48.55
10k obs, 1,500 epochs	43.1	44.1

A dropout rate of 10% is added to the first convolutional layer for both models, and the models are retrained. Both models show further improvement, and the discrepancy in performance between the two models is marginal. See Figure 3.11 and Table 3.4.

Figure 3.11: ResNet versus ResNext, 10,000 Observations for 1,500 Epochs with 10% Dropout Rate in First Convolution Layer

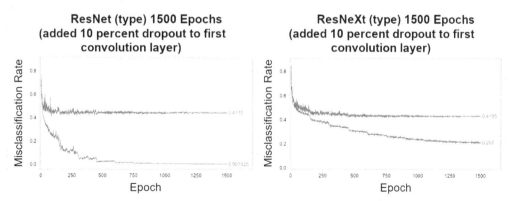

Table 3.4: ResNet versus ResNext, 10,000 Observations for 1,500 Epochs with 10% Dropout Rate in First Convolution Layer

Validation Misc	ResNet	ResNext
10k obs, 60 epochs	44.2	48.55
10k obs, 1,500 epochs	43.1	44.1
10k obs, 1,500 epochs, 10% dropout added	41.75	41.95

The data is increased from 10,000 observations to 50,000 observations. Variance for both models decreases, and the performance on the holdout data has drastically improved. The learning rate reductions are now clearly visible in the iteration plots. See Figure 3.12 and Table 3.5.

Figure 3.12: ResNet versus ResNext, 50,000 Observations, 1,500 Epochs

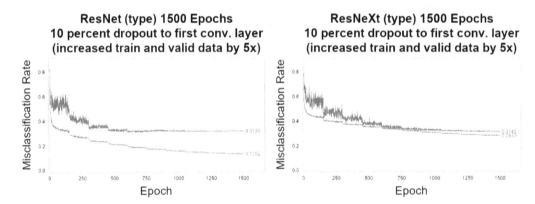

Table 3.5: ResNet versus ResNext, 50,000 Observations, 1,500 Epochs

Validation Misc	ResNet	ResNext
10k obs, 60 epochs	44.2	48.55
10k obs, 1,500 epochs	43.1	44.1
10k obs, 1,500 epochs, 100% dropout added	41.75	41.95

Validation Misc	ResNet	ResNext
50k obs, 1,500 epochs	31.26	31.45

Image Preprocessing and Data Enrichment

If there appears to be a significant difference in performance between the training and validation data partitions, then chances are that your model is suffering from high variance. Gathering or generating more data and applying regularizations can mitigate variance error. When generating images from the current data, translations and minor distortions are added to the new permutations. However, the translations should be random to avoid introducing bias. Enriching image data should be considered when there is limited training data and there is a large discrepancy between training and validation errors. This enrichment can be used to balance extreme target distributions.

Synthetic case creation is a practice used to enrich training data with **designer** cases when obtaining more data is infeasible. These artificial observations should represent possibilities that might exist in the population but might not be observed in the actual training sample. The key here is to create cases that represent the possible, **not the impossible.** For example, suppose that we are analyzing images of people exiting a train onto a station platform. We might want to add **reasonable** color distortions that might mimic changes in lighting. However, we would not want to vertically invert the picture because of gravitational influences present in the environment.

Sometimes, adding noise can improve model generalities. Consider modeling human speech for a new application. It might be beneficial to add in white noise to simulate the unpredictable environments in which the user might be attempting to use the speech application. Even adding random patches can also be considered a valid method used to generate more image data.

Data Augmentation Techniques

Data augmentation techniques often fall into one of four broad categories: geometric, photometric, domain, and generative.

- *Geometric transformations* involve changes in pixel orientation or position and are used to make the model more robust to spatial differences. This means that the model is more invariant to positional deviations of the elements in an image. Common transformations that fit into this category include rotating, flipping, scaling, and cropping.

- *Photometric morphologies and transformations* change the pixel intensity with the intended consequence of making the model more robust (or invariant) to changes in color. Common augmentation techniques include pixel inversion, lightening / darkening, color shifting, color jittering, eroding, and many others.

- *Domain transformations* incorporate data alterations driven by the modeler's knowledge of the data and the problem. Domain transformations usually incorporate a photometric or geometric transformation (or both). For example, a government agency building a facial recognition software using driver license photos might want to create additional training images that incorporate black circles around the individual's eyes to simulate sunglasses. These types of transformations are user generated.

- *Generative transformations* often rely on a generative model to create new translations of the input. These translations are defined by the model and a set of input variable distributions. The generative modeling field seems to be converging around one type of generative model, *generative adversarial models* (GANs) (Goodfellow et al. 2014). The primary downside to a generative model is the time that it takes to train and tune the model. Nevertheless, the models' generative capabilities are very impressive—so much so that, in some cases, the generated content can be indistinguishable to the human eye.

Figure 3.13: Examples of Image Transformations

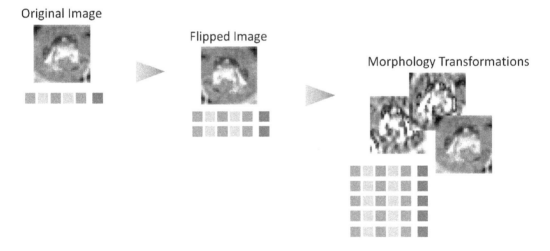

With all the complexities that accompany image data, generic synthetic case creation is (perhaps surprisingly) quite simple. That is, an image can be augmented with different techniques, such as flipping vertically, flipping horizontally, scaling up and down with pyramids, changing the contrast (darkening or lightening), sharpening, and rotating to the left or right. All these mutations can be applied to new copies of a training image to generate new cases.

More complex data augmentation techniques require extensive input from the modeler or incurring the expense of time used to train a generative adversarial model.

It is important to avoid mutations that might misrepresent the target. For example, vertically flipping a 6 turns it into a 9, but with the label of a six. Care should be taken to avoid these mistakes because they can degrade your model's performance.

The SAS Image action set contains a rich set of actions for image preprocessing and data augmentation. The Image action set offers the ability to load, process, and save images. It enables images to be used as a basic data type in SAS Viya. The load and save actions support all common image formats such as JPG for photographic images, and Digital Imaging and Communications in Medicine (DICOM) for biomedical images. The process action supports all images loaded from a photographic format. This chapter focuses on the ProcessImage and AugmentImage actions from the Image action set.

The ProcessImage and AugmentImage actions are often used for advanced augmentations. Augmenting the data before the modeling process incurs additional cost because the data size is increased, reducing the overall speed of the analysis. SAS deep learning provides basic augmentations "on the fly," which mitigates the speed reduction from data augmentation.

Process Image Action

The processImages action performs single or multiple image-processing functions on input images. The input of this action is a CAS table that contains images and an array of functions (with their specific parameters) as shown in the code below.

```
PROC CAS;
 ProcessImages /
        table= {name= "< image table name >"}
        casOut= {name= "< output image table name >"}

        imageFunctions= {
               {functionOptions= {functionType= "< first transformation
type >"                                    < function type-specific-
parameters >}},
               {functionOptions= {functionType= "< second transformation
type >"                                    < function type-specific-
parameters >}},
                       {{….}}};
RUN;
```

The output is a CAS table that contains the resulting images after the specified functions are applied. You can define the following functions:

- RESIZE: resizes an input image based on the width and height parameters. The parameters must be specified as positive integers.

- GET_PATCH: creates a patch based on a rectangular region (using x, y, width, and height parameters) from an input image. The x and y parameters must be specified as integers that are greater than or equal to 0. The width and height parameters must be specified as positive integers. Further, if x + width is greater than the specified width of the input image, then the width of the rectangular region is reduced to the input image's specified width. The same is true for the y and height parameters.

- CANNY_EDGE: detects edges in input images by using the Canny edge detection algorithm. This function accepts three parameters: lowThreshold, highThreshold, and kernelSize.

- LAPLACIAN: applies the Laplace operator to input images. The Laplace operator is a second-order differential operator in the n-dimensional Euclidean space. When the value of the kernelSize parameter is 1, the resulting value is approximated by convolving the kernel with input images.

- Otherwise, the value is approximated by summing the second x and y derivatives, which are calculated using the Sobel operator.

- SOBEL: uses the Sobel operator to calculate the first, second, third, or mixed image derivatives. Derivatives are calculated by convolving the image with a kernel. Kernels must be square. The size can be only one of the following numbers: 1, 3, 5, and 7. When the kernel size is 1, a kernel is used for the first or second derivative of x, and a kernel is used for the first or second derivative of y. For the other kernel sizes, the Sobel operator considers Gaussian smoothing and differentiation together (handling noise).

- NORMALIZE: normalizes the value range of an image. You can specify the following normalization functions in the type parameter: INF, L1, L2, L2SQR, HAMMING, HAMMING2, RELATIVE, and MINMAX. The MINMAX function is commonly used to normalize an image between alpha and beta values.

- THRESHOLD: applies a threshold value to each pixel in an image. You can specify the following threshold types: BINARY, BINARY_INVERSE, TRUNCATE, TO_ZERO, TO_ZERO_INVERSE, OTSU, and TRIANGLE.

- CONVERT_COLOR: converts an image's color space. Currently available conversions are COLOR2GRAY, GRAY2COLOR, BGR2RGB, and RGB2BGR. The default color space for an image is BGR. COLOR2GRAY means BGR2GRAY.

- RESCALE: changes an image's depth. You can specify the following types of rescaling:

  - TO_8U where depth is changed to 8-bit, where a pixel is represented by an unsigned integer.

  - TO_32F where depth is changed to 32-bit, where a pixel is represented by a floating number.

  - TO_64F where depth is changed to 64-bit, where a pixel is represented by a double-precision number).

  You can scale the values based on the alpha and beta parameters.

- MORPHOLOGY: performs a morphological transformation on images. You can specify the following types of transformation: ERODE, DILATE, OPEN, CLOSE, GRADIENT, TOPHAT, BLACKHAT, and HITMISS. Each of these operations can be in RECT (rectangular), CROSS, and ELLIPSE shapes.

- BOX_FILTER: blurs an image by using a normalized box filter. The kernelWidth and kernelHeight parameters shape the filter. The anchor is assumed to be at the kernel's center.

- GAUSSIAN_FILTER: blurs an image by using a Gaussian filter. A convolution kernel is created based on the kernelWidth and kernelHeight parameters, which must be positive and odd.

- BILATERAL_FILTER: applies a bilateral filter to images. A bilateral filter is a nonlinear, edge-preserving, noise-reducing smoothing filter. The idea behind this filter is to consider pixels close if they are
in spatially nearby locations, and similar if they have nearby values (in color space).

- MEDIAN_FILTER: blurs an image by using the median filter. This filter uses a square matrix of the size specified in the kernelSize parameter, which must be greater than 1 and odd.

- BUILD_PYRAMID: blurs an image and samples it down (PYR_DOWN) or up (PYR_UP). A Gaussian kernel is used for blurring.

- CONTOURS: applies several preprocessing steps for more accurate results. This function accepts only one-channel images (for example, grayscale images). It provides better results when the input image is a binarized image (for example, pixel values are either black or white). If you provide a grayscale image rather than a binarized image, you still see results, but they might be inaccurate or unexpected. (For example, a contour could look like a frame that contains the whole picture.)

- CUSTOM_FILTER: creates a custom filter and uses it in convolution. This operation takes the width, height, and values of a filter.

- ADD_CONSTANT: adds or subtracts a constant value from the pixel intensity values input images. This function also makes sure that the resulting values are in the permitted range that is determined by the depth of a pixel. For example, if an image's pixel depth is unsigned 8-bit (meaning that the pixel values can be between 0 and 255 only), then this function makes sure that the resulting values are within the range of 0 and 255 (meaning no overflows and negative values are permitted in this case). Furthermore, if the image is a one-channel image, then the function uses only the constant value of the first input.

- HIST_EQUALIZATION: aims to stretch out the intensity range. This function maps the intensity distribution of an input image to another distribution that is wider and has a more uniform distribution of intensity values. Two versions are supported: global and adaptive. In the global version, the mapping is applied by considering the whole image. In the adaptive version, the mapping is performed in local patches of input images. Therefore, the adaptive version is more robust to images where contrast is quite different at different regions of those images.

- LINEAR_TRANSFORMATION: performs linear transformation in one of the following ways, based on the value of the method parameter:

  - STANDARDIZATION creates an image that has zero mean and unit variance.

  - WHITENING_PCA uses principal component analysis to create an image that has an identity covariance.

  - WHITENING_ZCA uses a ZCA transformation (also called a Mahalanobis transformation) to create an image that has an identity covariance.

  - LOCAL_CONTRAST_NORM is similar to STANDARDIZATION except that it convolves a 9 x 9 Gaussian image with the input image.

- MUTATIONS: mutates images by using different augmentation techniques: flipping vertically, flipping horizontally, scaling up and down with pyramids, changing the contrast (darkening or lightening), sharpening, and rotating to the left or right. This function is useful for increasing the variety of the input images, which is a major preprocessing step for training methods based on deep neural networks.

The code below displays an example code that demonstrates how to use the processImage action to apply a 5 x 5 Gaussian filter to blur an image. The KERNELWIDTH= and KERNELHEIGHT= options specify the width of the kernel width and height, respectively.

```
PROC CAS ;
 ProcessImages /
       table={name= "MyInputTable"}
       casOut={name= "MyOutputTable"}

       imageFunctions= {
             {functionOptions= {functionType= "Gaussian_filter"
                         KernelWidth= 5
                   KernelHeight= 5}};
RUN;
```

The augmentImages action enables you to create patches by using either sliding windows or coordinates of a rectangle. It also allows image augmentation by using a number of methods, including rotating the image, flipping it, and adjusting the contrast as seen in the code below.

```
PROC CAS;
 AugmentImages /
       table={name= "< image table name >" }
       casOut={name= "< output image table name >" }
       writeRandomly= < true | false
             cropList= {mutations= { < first mutation type > = < true |
false >
                                   < second mutation type > = < true| false
>
                                   < ... > }
                   width= < input image width >
                   height= < input image height >
                   useWholeImage= true | false};
RUN;
```

This action can process more than one patch or augmentation command in a single run.

For example, you can use the sliding window method on input images and separately flip input images. The output CAS table contains the resulting images. In addition, the writeRandomly parameter enables you to randomly apply augmentations to input images. This is very important for machine learning methods that are based on deep neural networks.

Setting both the sweepImage and useWholeImage parameters to True is contradictory. Therefore, the useWholeImage parameter overrides the sweepImage parameter.

The code below is an example that demonstrates how to use the AugmentImage action to apply two transformations to a 32 x 32-pixel image.

```
PROC CAS;
 AugmentImages /
       table={name= "MyInputTable"}
       casOut={name= "MyOutputTable"}
       writeRandomly= true

          cropList={mutations= {      colorjittering= true
                                rotateright= true            }
                    width= 32
                    heigth= 32
                    useWholeImage= true};
RUN;
```

Gaussian Filters

Gaussian filters are used to smooth steep variations within an image.

Figure 3.14: Gaussian Filter

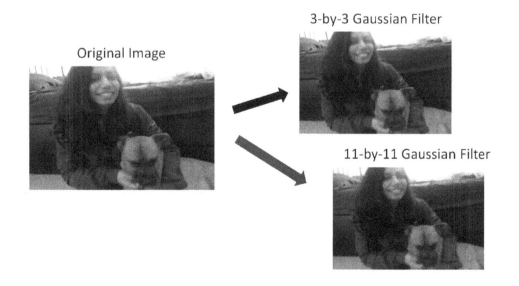

This photometric transformation blends pixels to create a blurred version of the image. Larger kernels increase the blurred effect of the transformation as seen in Figure 3.14. A Gaussian filter applies larger weights to the center of the kernel such that the area of interest is lightly blended with surrounding pixels. An example Gaussian kernel filter might look something like the figure below.

Figure 3.15: Gaussian Kernel Filter

1	2	1	1	1
1	5	4	6	1
2	7	14	8	1
3	5	6	5	2
2	1	2	3	2

The user needs to provide only the height and width of the Gaussian kernel. SAS creates a *valid* kernel such that the volume under the two-dimensional kernel equals 1. The processImages action is used to create images transformed by Gaussian filters. Here is some example code:

```
PROC CAS;
   image.processImages /
         imageFunctions= {{functionOptions=
      {functionType= "GAUSSIAN_FILTER"

      KernelWidth= 5
            KernelHeight= 5}}
         table={name='inputTable'}
         casout={name='outputTable'};
RUN;
```

Sharpen

Sharpening an image is considered a photometric transformation, as seen in Figure 3.16.

Figure 3.16: Sharpen

Original Image

To sharpen an image, SAS encodes the red, green, blue (RGB) color image to Y'CbCr, where Y' is the brightness component, Cb represents the blue-difference component, and Cr represents the red-difference component. A *high-pass* filter is then applied to the encoded image to detect edges. Figure 3.17 is an example of a high-pass filter.

Figure 3.17: High-Pass Filter

0	-1	0
-1	5	-1
0	-1	0

Then the image is transformed back the RGB color space. Here is example code that sharpens an image:

```
PROC CAS;
   image.processImages /
         imagefunctions= {{functionOptions=
      {functionType="MUTATIONS"

      type="SHARPEN"
         }}}
         table={name='inputTable'}
         casout={name='outputTable'};
RUN;
```

Inverting Pixels

Pixel inversion is a photometric transformation that reverses the color density such that light becomes dark and dark becomes light, as seen in Figure 3.18.

Figure 3.18: Inverting Pixels

It can be as simple as replacing each of the RGB pixel values with (255 – pixel value). Here is sample code that demonstrates how to invert pixels using SAS:

```
PROC CAS;
   image.processImages /

        imagefunctions= {{functionOptions=
        {functionType="MUTATIONS"

        type="INVERT_PIXELS"
        }}}
        table={name='inputTable'}
        casout={name='outputTable'};
RUN;
```

Pyramid Down

The Pyramid Down transformation downsamples and then blurs the image. An example is shown in Figure 3.19.

Figure 3.19: Pyramid Down and Cropping Transformations

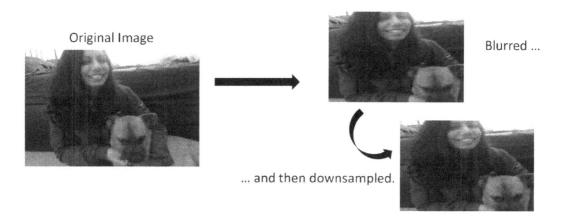

Cropping was also used in the example in Figure 3.19. Empirical evidence indicates that cropping, a geometric augmentation, can significantly improve the generalization performance of a convolutional neural network model (Taylor and Nitschke 2017). The augmentImages action is used to both crop and apply the Pyramid Down transformation in the same pass.

In the code below, notice that the useWholeImage= option is set to **false**, which indicates cropping will likely be performed.

```
proc cas;
    image.augmentImages   /
      cropList= {{mutations=
{pyramidDown=TRUE}
x=60
y=20

width=416
height=416

outputWidth=208
outputHeight=208

useWholeImage=false
        }}
      table={name='inputTable'}
      casout={name='outputTable'};
run;
```

The X= and Y= options represent the pixel starting position. The WIDTH= and HEIGHT= options specify the original image size. The OUTPUTWIDTH= and OUTPUTHEIGHT= options control the extent to which the image will be downsampled. In the example above, the image is down sampled by a factor of two (416 / 208).

Rotating

Rotating an image is one of the most popular geometric transformations used to expand training data for computer vision tasks because an object's angle can often vary in application. Consider an image of a person sitting upright in a chair versus a person reclined back in a chair. Reclining back in the chair alters the angles of vital facial features for which the model might be relying on to detect the person. Figure 3.20 shows ways an image can be rotated.

Figure 3.20: Image Rotation

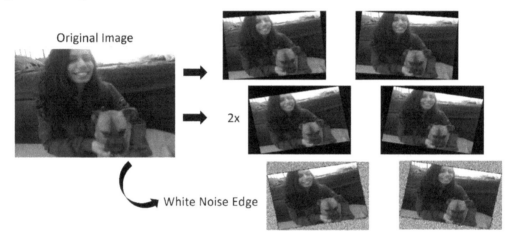

An image can be rotated either left or right, and the transformation can be applied multiple times to the same image. SAS automatically pads the outer edges with zero-padding. However, replacing the zero-padding with noise might help the model's generalizability. Alternatively, you can rotate and then crop the image to remove the zero-padding region. Here is code that rotates an image to the right using the processImages action:

```
proc cas;
   image.processImages /

        imagefunctions= {{functionOptions=
        {functionType="MUTATIONS"

        type="ROTATE_RIGHT"
        }}}
        table={name='inputTable'}
        casout={name='outputTable'};
run;
```

Flipping

Flipping an image is another widespread geometric transformation, but it is one that should be used with care because it can sometimes create a mislabeled image or represent an unlikely scenario. Figure 3.21 shows an example of flipping an image.

Figure 3.21: Flipping

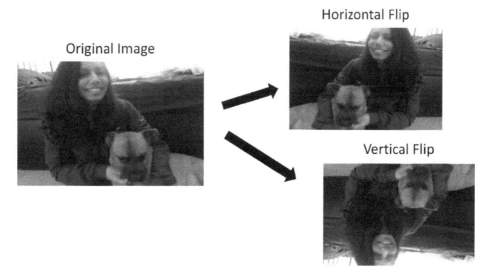

The code below flips an image vertically using the processImages action.

```
PROC CAS;
   image.processImages /

        imagefunctions={{functionOptions=
        {functionType="MUTATIONS"

        type="VERTICAL_FLIP"
        }}}
        table={name='inputTable'}
        casout={name='outputTable'};
run;
```

In most cases, multiple photometric and geometric data augmentation techniques are randomly applied to images to generate a diverse set of training examples, as seen in Figure 3.22.

Figure 3.22: Multiple Data Augmentation Techniques

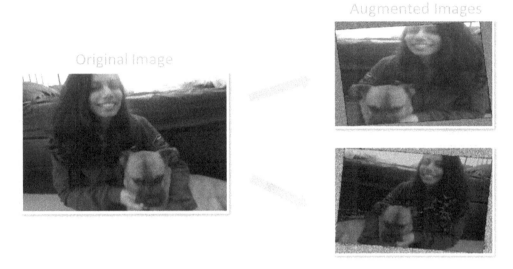

Data augmentation techniques are usually not applied to validation or test partitions to reduce the pressures of overfitting.

Transfer Learning Introduction

Traditional machine learning methods extract information from one data source for use in another data source under the assumptions that the distributions are the same, or at least very similar, as in Figure 3.23.

Figure 3.23: Similar Distributions

Predictive modeling is a classic example where information extracted from historical or current data is used to predict an outcome associated with data not yet observed. Predictive models assume the feature distributions remain the same from training to scoring.

Transfer learning is an area of machine learning that leverages information extracted from one set of distributions for use in another different set of distributions as seen in Figure 3.24.

Figure 3.24: Dissimilar Distributions

That is, transfer learning refers to a situation where information extracted from one setting (often referred to as the source) is exploited in another setting (often referred to as the target) to improve the learner's performance. Meaning the feature space of the extracted information is different from the feature space associated with the desired learner (model). Transfer learning refers to a situation where information extracted from one setting is exploited in another setting with the goal of improving the learner's performance. More formally, either the source (*s*) and target (*t*) *domains* do not match ($D_s \neq D_t$) or the source and target task do not match ($T_s \neq T_t$) (Weiss, Khoshgoftaar, and Wang 2016).

Domains and Subdomains

A *domain* is an environment or setting that is described by a set of features. (See Figure 3.25.) Domains are antecedents to subdomains, where the structure of domains and subdomains describe a hierarchical relationship of related information. Domains, and by extensions subdomains, are defined by a feature space (*X*) and a marginal probability distribution (*P* (*X*)). For example, English, Chinese, and Icelandic translations are each subdomains of a natural language processing domain (Weiss, Khoshgoftaar, and Wang 2016).

Figure 3.25: Example of Domains and Subdomains

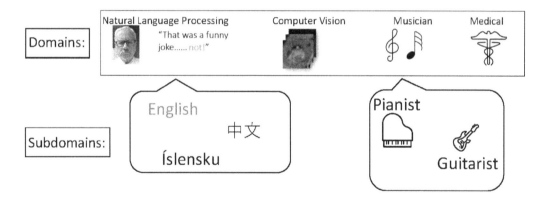

> **Note:** Usually, the nearest subdomain is considered as the "domain" for practical purposes when considering whether the source and target domains are equivalent. That is, when answering the question is $X_S = X_T$ and $P(X_S) = P(X_T)$?

Tasks

Domains consider only the input variables, whereas the task, on the other hand, considers the outcome. Tasks are defined by an output space (*Y*) and a conditional probability distribution (*P(Y|X)*). For example, Figure 3.26 shows that predicting a classification outcome, such as the breed of a dog based on an image of the dog, is a task within the computer vision domain (Weiss, Khoshgoftaar, and Wang 2016).

Figure 3.26: Domain versus Task

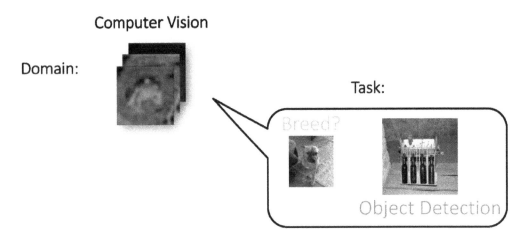

Therefore, we can formally define *transfer learning* as, "given a source domain D_s with a corresponding task T_s and a target domain T_s with a corresponding task T_t, transfer learning is the process of improving the target function $f_t(x)$ $f_t(x)$ by using the related information from D_s and D_t, where $D_s{\neq}D_t$ or $T_s{\neq}T_t$" (Weiss, Khoshgoftaar, and Wang 2016).

Types of Transfer Learning

There are two types of transfer learning: homogeneous (where $X_s=X_t$) and heterogeneous (where $X_s{\neq}X_t$). An example of homogeneous transfer learning is shown in Figure 3.27 below, where each column represents a pixel density value.

Figure 3.27: Homogeneous Transfer Learning

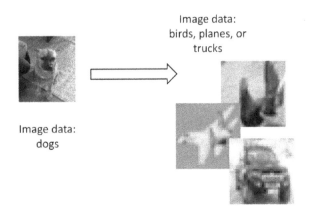

> **Note**: $P(X_s)$ does not have to equal $P(X_t)$ in homogenous transfer learning.

Heterogenous Transfer Learning and Domain Adaptation

As mentioned above, heterogenous transfer learning refers to cases where $X_s \neq X_t$. An example of heterogeneous transfer learning is shown in Figure 3.28.

Figure 3.28: Heterogeneous Transfer Learning with Tabular and Image Data

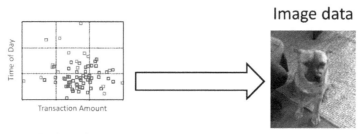

Domain adaption is closely related to heterogeneous transfer learning. It is characterized by methods that alter the source domain (X_s and/or $P(X_s)$) to better fit the target domain in situations where the source and target task are the same $T_s = T_t$, but the domains are different $D_s \neq D_t$. For example, consider a task of text generation, which consists of generating text based on a set of criteria. An example of criteria could be the demeanor (happy, sad, angry, and so on) of the generated text. The domains are different when generating text for a Chinese conversation, as compared to an English conversation. Both tasks are the same, but the vocabulary, style, linearity, and systematicity of the languages differ. Domain adaption would alter (or adapt) the source domain to match the target domain. Glorot et al. (2011) demonstrated stacked denoising autoencoders could be used for domain adaption.

> **Note**: There are many inconsistencies in literature with regard to the formal definition of domain adaption. This book's definition of domain adaptation is consistent with both Goodfellow et al. 2016, and Weiss et al. 2016.

Transfer Learning Biases

Transfer learning can be difficult because by invention, there exists a difference in the information being transferred from one setting to another. These differences can give rise to biases and can cause the model to degrade to the point of *negative transfer learning*.

Negative transfer learning occurs when the desired target function performs worse after having incorporated a transfer learning strategy relative to the target function with no exposure to transfer learning. Bias is the primary cause of negative transfer learning. The two most common types of transfer learning biases are *frequency bias* and *context bias*.

Frequency Bias

Frequency bias occurs when features are not independent of the domain, and the transfer learning strategy is being deployed across different domains. For example, consider the task of sentiment analysis across two domains, a hospital domain, and a business domain. Words such as "people," "help," and "work," are independent of the domains and are likely to occur at similar rates. However, words such as "heart," "patient," and "medication" are likely to occur more frequently in the hospital domain, and words like "profit," "customer," and "sales" would be more frequent in the business domain. Therefore, $P(X_s) \neq P(X_t)$, meaning the marginal probability distribution of the features does not match.

Context bias occurs when the underlying relationship between the features and the output changes from source data to target data. Consider our earlier example of the hospital and business domains. The word "positive" used in a business domain would likely have positive sentiment. However, that same word used in a hospital domain might have negative sentiment.

For example, "the test results came back positive," where the word "positive" is in reference to the presence of a disease. Therefore, context bias refers to instances where $P(Y_s \mid X_s) \neq P(Y_t \mid X_t)$, meaning that the conditional probability distributions are different from the source data to the target data.

Transfer Learning Strategies

There are many methods that are used for transfer learning. Two of the most general forms are supervised and unsupervised pretraining. This chapter focuses on specific examples of supervised and unsupervised pretraining that can be used to capture and leverage information that exist across domains. Transfer learning is also known as *representational learning*.

Supervised Pretraining

One method for supervised pretraining incorporates first training a model on the source data. The model weights are then transferred and frozen for use by the target data. It is common to leave some weights unfrozen to allow for reconciliation between the weights learned on the source data, and differences in the target data. In neural network modeling, the layers with the unfrozen weights are referred to as the "adaptable layers." The output layer is left unfrozen, permitting the generation of new posteriors that reflect the target data.

Choosing which layers to freeze depends on the problem at hand. If the task is identical from the source to the target ($Y_s = Y_t$), then sharing the layers closer to the output layer is sensible. In this scenario, the adaptable layers are those closer to the input layer. Conversely, the layers closer to the input layer should be frozen and the layers closer to the output left adaptable if the relevant information exists primarily in the inputs.

Unsupervised Pretraining

Autoencoders are unsupervised models that have been leveraged with some success for unsupervised pretraining. One benefit to an unsupervised method is that the source data do not need to be labeled. An unsupervised pretraining strategy usually begins by combining the features of the source and target data. An autoencoder is then trained on the full data with hopes of capturing information that persists across domains. This is most useful when the source data is abundant and the target data is limited. The trained autoencoder is then assigned to score (inference) the target data and encoded projections are extracted. These encoders are provided as inputs to the target learner, along with the original target features. Glorot et al. (2011) have shown success using stacked denoising autoencoders for unsupervised transfer learning.

Denoising autoencoders were introduced by Vincent et al. (2010) as autoencoders that are trained on corrupted (perturbed) inputs. Corrupting the inputs can be as simple as applying dropout to the input layer. A popular variant of a denoising autoencoder is called a *stacked denoising autoencoder*.

A stacked denoising autoencoder is greedily trained, layer by layer. Meaning, the first layer is thoroughly trained on the perturbed inputs and then frozen. The perturbed input is then replaced with the unperturbed input and the next layer is trained thoroughly. This layer is then frozen, and the process repeats until the output layer is trained.

Another variant of autoencoders is known as *sparse autoencoders*. Sparse autoencoders incorporate a constraint on the objective function. L1 and L2 regularizations can be used as the constraining mechanisms to generate a sparse autoencoder.

Customizations with FCMP

Sometimes data scientists want to create custom functions for models to better fit a particular deep learning requirement that the available default functions cannot deliver. Deep learning tools in SAS enable you to use the SAS Function Compiler (FCMP) to create and modify your own custom definitions for deep neural network activation and error functions. FCMP is a programming tool that you can use with the deep learning actions in SAS to produce your own custom deep learning entities:

- activation functions
- error functions
- model layers
- learning rate policies

Custom FCMP network layers, functions, and learning rate policies are separate entities that perform different deep learning tasks. The one common factor is that all of the custom layers, custom activation and error functions, and custom learning rate policies were created for use with SAS deep learning using FCMP.

The parameters of the FCMP action can be found in the documentation.

Tuning a Deep Learning Model

Selecting Hyperparameters

The quality of the predictive model that a machine learning algorithm creates depends on the values of various attributes that govern the training process and model structure. The parameters that represent these attributes are also known as *hyperparameters*. Searching the hyperparameter space of a deep neural network can be prohibitively expensive for genetic or Bayesian optimization methods, because these methods require training with each hyperparameter combination until convergence of the algorithm. Alternatively, the Hyperband method (Li et al. 2017) is a good approach for dense model structures, because resources are adaptively allocated to only those model structures that show promise.

Sampling the Hyperparameter Space

The dlTune action implements the Hyperband method and begins by selecting sets of hyperparameters using a Latin hypercube sample of the search space. A Latin hypercube sample is preferred over a simple random, which is considered the least intelligent sample search. (See Figure 3.29.)

Figure 3.29: Simple Random Sample

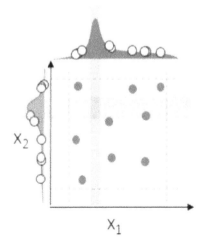

The *Latin hypercube sample* is a discrete sample that is uniform in each hyperparameter but random in combinations. These so-called *low-discrepancy point sets* attempt to ensure that points are approximately equidistant from one another in order to fill the space efficiently, as seen in Figure 3.30.

Figure 3.30: Latin Hypercube Sample

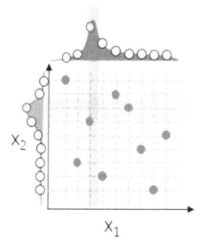

This sampling allows for convergence across the entire search range of each hyperparameter. It is more likely to find good values of each hyperparameter when compared to a simple random sample. The method is considered to yield a fairly uniform sampling of the space (Koch, et al. 2018).

To avoid overfitting, a holdout sample is recommended to evaluate each set of hyperparameters. Each set of sampled hyperparameters is assigned to the specified model structure, and resources are allocated proportionally per the Hyperband method, as seen in Figure 3.31.

Figure 3.31: Hyperband Tuning

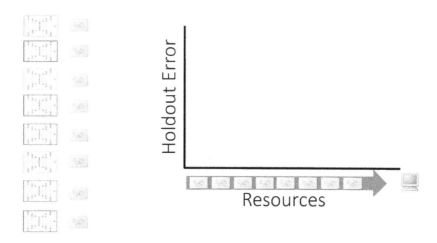

The models are then trained for one or several epochs, and their accuracy assessed. Poor performing models are then removed from the training process. The user can control the proportion of hyperparameter sets that are retained with the TUNERETENTION= option. See Figure 3.32.

Figure 3.32: Hyperband Tuning Over One or Several Epochs

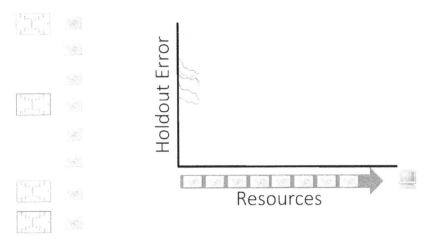

The resources previously provided to the poor performing models are reallocated and distributed among the remaining models, as seen in Figure 3.33.

Figure 3.33: Poor Performing Model Resource Reallocation

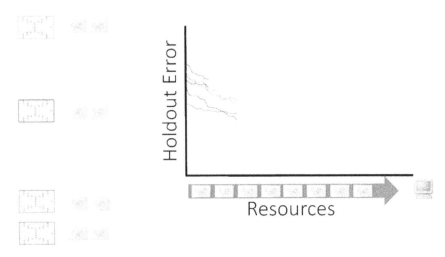

The remaining models are then trained for one or several epochs, and their accuracy is assessed. The process is repeated until a champion set of hyperparameters remains (Figure 3.34).

Figure 3.34: Champion Model

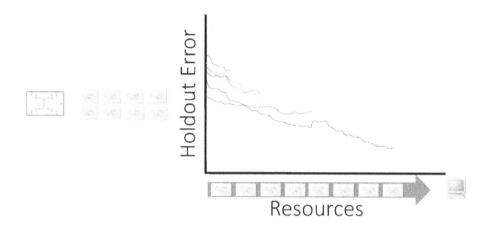

Hyperparameter Selection

You can tune the following hyperparameter values through DLTUNE:

- Learning rate
- Momentum rate
- Mini-batch size
- Learning rate reduction (gamma)
- Step size
- L1 Regularization
- L2 Regularization
- Dropout
- Beta1
- Beta2

The dlTune action can search for optimal values amongst many important hyperparameters. Tuning the learning rate (or rates) is recommended for easy gains in model lift. Tuning the regularizations (L1, L2, and dropout) is recommended if the model performs significantly better on the training data than on the validation data.

> **Note:** When the gpu parameter is specified, the dlTune action uses graphical processing unit hardware to accelerate the dlTune action. You can identify GPU devices with 0-based 64-bit integers. For example, devices={0, 2} requests that the first and the third GPU device be used. When the devices= option is not specified, the action uses all available GPU devices in the system.

Hyperband Properties

The following optimization property values affect the behavior of DLTUNE:

- MAXEPOCHS (Maximum epochs)
- TUNEITER (Tuning iterations)
- NUMTRIALS (Number of trials)
- TUNERETENTION
- TUNERESTART

The MAXEPOCHS property specifies the number of epochs to perform on each hyperparameter combination before comparing validation errors and culling. The number of combinations culled is a function of the TUNERETENTION property.

> **Note:** For SGD with a single-machine server or a session that uses one worker on a distributed server, one epoch is reached when the action passes through the data one time. For a session that uses more than one worker, one epoch is reached when all the workers exchange the weights with the controller one time. The syncFreq parameter specifies the number of times each worker passes through the data before exchanging weights with the controller.

The TUNEITER property specifies the number of iterations of the tuning algorithm to run. One iteration of the tuning algorithm trains the models using all active hyperparameter sets for the specified maxEpochs. The hyperparameters with the best validation fit error remain active for future tuning iterations.

The NUMTRIALS property specifies the number of hyperparameter sets to try at the start of the tuning process. For example, specifying 100 for the property will begin the process training 100 models with different hyperparameter values.

The TUNERETENTION property determines the proportion of hyperparameter configurations that are kept after each tuning iteration.

The TUNERESTART specifies the number of times to restart the parameter tuning process, using the best available weights from a previous search, and newly chosen parameter sets.

Choosing the appropriate set of tuner values for maxepochs, tuneiter, numtrials, tuneretention, and tunerestart is not trivial. A common approach is to set the maxepochs number to a small value so that the number of hyperparameter combinations (numtrials) can be assigned a larger value. This approach seems to imply that for hyperparameter search methods, the model parameter estimates (weights) matter less than the characteristics of a model. Regardless of whether this is true or not, there are instances when an alternative approach is warranted. For example, imagine your model's current iteration plot resembles Figure 3.35.

Figure 3.35: Current Iteration Plot

It would then be logical to search L1 and L2 regularization values in hopes of reducing variance in the model. Setting the maxepochs value to a small number, such as four, might result in the process assessing the performance of each hyperparameter combination at a peak of poor performance, as seen in Figure 3.36.

Figure 3.36: Setting Maxepochs to a Small Value

Assessing the model at this point in the training process is likely to favor larger L1 and L2 values because larger values constrain weight growth, which is likely to cause the model to be less responsive to information in the data. Meaning, larger L1 and L2 values appear to have superior performance at epoch four because the degradation observed early in the process will be dampened. But are large L1 and L2 weight values really "best" for the model in the long run? One might postulate the answer is "no" because the model will struggle to capture less pronounced single that might exist in the data. Therefore, in this example it may be best to set maxepochs to a larger value to avoid assessing the hypermeter combination at the poor performance peak observed early in the training process, as seen in Figure 3.37.

Figure 3.37: Setting Maxepochs to a Larger Value

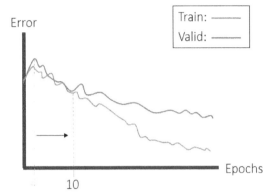

It is imperative that the business problem and the model's interaction with the data should be considered when setting the tuning parameters.

Chapter 4: Object Detection

Introduction

By itself, *image classification* is a task of examining an image that is usually well framed and responding with a probability of classification that describes the image. *Object detection* is more advanced than image classification because object detection analyzes complex images that contain a mixed multitude of objects, at different distances and locations, amidst varying, often visually noisy backgrounds. Objects can appear anywhere within the visual frame, be near or far, and can overlap each other. (See Figure 4.1.) Object detection locates and classifies objects, and it determines these objects' boundaries and relationships to other objects.

Figure 4.1: Object Classification

Object detection is a challenging task, as well as one of the most fundamental tasks in computer vision. We use convolutional neural networks (CNNs) to perform tasks that include object identification, object localization, and bounding, as well as object classification. Lately, CNN-based deep learning algorithms like the following have been successful in addressing problems associated with object detection tasks:

- You Only Look Once (YOLO)
- Single Shot MultiBox Detector (SSD)
- Region Proposal Networks (R-CNN)
- Faster Region Proposal Networks (Faster R-CNN)
- Backbone feature extractor for object detection (DETNet)

Types of Object Detection Algorithms

Object detection algorithms can be categorized as follows. The first category of object detection algorithms looks for objects in one step with anchors of predefined sizes at predefined locations. The locations and fixed sizes are strategically selected in order to cover the greatest number of scenarios. The algorithms typically separate the original images into fixed-size grid regions. For each region, these algorithms try to predict a fixed number of objects using certain predetermined shapes and sizes. Algorithms in this category are called *single-stage methods*. Examples of single-stage algorithms include YOLO, SSD, and RetinaNet. Algorithms in this category usually run faster but can be less accurate. This type of algorithm is often used for applications that require real-time detection.

The second category of object detection algorithms is region proposals that involve two steps. Under region proposal, the regions that are highly likely to contain an object are selected either using traditional computer vision techniques (like selective search) or by using a deep-learning-based region proposal network (RPN). After you gather the small set of candidate windows, you can formulate a specific number of regression models and classification models to solve the object detection problem. This category includes algorithms like Faster R-CNN, R-FCN, and FPN-FRCN. Algorithms in this category are usually called *two-stage methods*. They are generally more accurate but are often slower than single-stage methods.

The deep learning tools in SAS support two representative object detection algorithms: YOLO and Faster R-CNN. YOLO is a one-stage, fixed-size object detection algorithm. Faster R-CNN is a two-stage region, proposal-based object detection algorithm.

One-stage object detectors like YOLO analyze an image in a single pass, and they output multiple object location and classification predictions. As a result, YOLO networks are fast when compared to multi-stage algorithms like R-CNNs. YOLO architectures are also able to perform image reasoning with larger contexts during training.

YOLO applies a single CNN to the full image and then divides the image into regions (usually 13 x 13 grid cells). Then, for each region, YOLO predicts a fixed number of bounding boxes and associated object classification probabilities. Anchor boxes for each region provide pre-set values for predicted object sizes, and the actual prediction gives a correction to the anchor boxes. Anchor boxes are typically obtained for a given data set using k-means clustering. Final detection results are obtained by applying threshold and non-maximum suppression operations on the predictive probabilities.

Data Preparation and Prediction Overview

The data from which the model now learns has expanded to include information describing the spatial dynamics of an object or objects. Depending on the format used, the predictions that describe an object's location can be presented in the following three ways:

1. RECT: Use the RECT format for coordinates [xleft, ytop, width, height] in image pixels
2. YOLO: Use the YOLO format for coordinates [xMiddle, yMiddle, width, height] in normalized image size (obtained by dividing the coordinates by the image size)
3. COCO: Use the COCO format for coordinates [xmin, ymin, xmax, ymax] in image pixels

An output variable describing the number of objects is included in addition to the output variables describing an object's location. This chapter focuses on the YOLO format, which includes four normalized variables that depict an object's location. Normalizing the values ensures that resizing transformations can be used with minimal loss of precision with regard to the target location.

- **_Objectn_x** corresponds to the normalized horizontal location of the object.

- **_Objectn_y** corresponds to the normalized vertical location of the object.

- **_Objectn_width** corresponds to the normalized width of the bounding box.

- **_Objectn_height** corresponds to the normalized height of the bounding box.

Non-Normalized Data

In some cases, the data provided are not normalized. For example, consider a scenario where xmin, max, ymin, and ymax represent unnormalized pixel values in an image with dimensions of 500 x 400. The normalized values are calculated by dividing each x and y coordinate by the maximum value of the respective plane (that is, xmin/500, xmax/500, ymin/400, and ymax/400).

The following equations transform the normalized xmin, xmax, ymin, and ymax values into YOLO formatted values:

_objectn_x=.5(x min + x max)

_objectn_y=.5(y min + y max)

_objectn_width=(x max - x min)

_objectn_height=(y max - y min)

Where *n* represents the object number.

Normalized Locations

Consider the example below where the object of interest is a person fishing in Figure 4.2. Pixels within a two-dimensional image represent coordinates on an X-axis and Y-axis grid. The grid itself can be thought of as a normalized representation of the image where the pixel in the top left corner has the coordinates (0,0), and the pixel in the bottom right corner has the coordinates (1,1). The variables **_objectn_x** and **_objectn_y** provide the normalized origin of an object. For example, the fisherman is located approximately 70.5% in the direction of the X axis' terminal value, 1.

Figure 4.2: Normalized Locations, objectn_x and objectn_y

ImageID	_Object0_	_object0_x	_object0_y	_object0_width	_object0_height
00006b13c052138f	Person	0.70594	0.73077	0.098125	0.23452

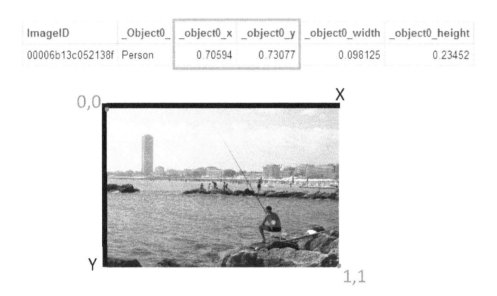

This means that the fisherman's origin is 70.5% of the way to the right side of the image. Similarly, the fisherman is located 73% in the direction of the Y axis' terminal value (that is, 73% of the way heading from top to bottom). The combination of _objectn_x and _objectn_y provides the fisherman's normalized origin.

The variables _objectn_width and _objectn_height provide the respective normalized width and height of an object, as seen in Figure 4.3. The object's origin (in this case, the fisherman's origin) lies in the center of the box. This means that the width and height extend in equal distances from the fisherman's origin.

Figure 4.3: Normalized Locations, objectn_width and objectn_height

ImageID	_Object0_	_object0_x	_object0_y	_object0_width	_object0_height
00006b13c052138f	Person	0.70594	0.73077	0.098125	0.23452

Notice that the width of the box is approximately 10% the width of the image and the height of the bounding box is about one quarter, or 23%, the height of the picture.

Multi-Loss Error Function

The YOLOv2 multi-loss function combines several error functions and attempts to balance the model's focus between classification and localization prediction task. The prediction task focuses on the following four areas:

- bounding box error $\lambda_{coord} \sum_{i=0}^{S^2} \sum_{j=0}^{B} L_{ij}^{obj} [(z_x^{ij} - \hat{z}_x^l)^2 + (z_y^{ij} - \hat{z}_y^l)^2 + (z_w^{ij} - \hat{z}_w^l)^2 + (z_h^{ij} - \hat{z}_h^l)^2]$

- $+$ true confidence error $\lambda_{obj} \sum_{i=0}^{S^2} \sum_{j=0}^{B} L_{ij}^{obj} (C_{ij} - 1)^2$

- $+$ false confidence error $\lambda_{noobj} \sum_{i=0}^{S^2} \sum_{j=0}^{B} L_{ij}^{noobj} (C_{ij} - 0)^2$

- $+$ classification error $\lambda_{class} \sum_{i=0}^{S^2} L_{ij}^{obj} \sum_{c \in classes} (p_{ij}(c) - \hat{p}^l(c))^2$

As seen in Figure 4.4, the image is separated into a matrix of S x S cells. In each cell, a number B of bounding boxes are predicted. Each prediction includes the center location of the bounding box and the width and height of the bounding box error, the probability of the prediction being an object (source of true and false confidence errors), and the classification probabilities (source of classification error). The number of bounding boxes predicted are determined by the number of anchor boxes specified. Anchor boxes are discussed later in the chapter.

L_{ij}^{obj} is 1 if the prediction j in the cell i is correct for a labeled object whose center falls into the cell i. Otherwise, its value is 0. L_{ij}^{noobj} is 1 if the prediction j in the cell i is **not** correct for a labeled object whose center falls into the cell i. Otherwise, its value is 0.

Figure 4.4: YOLOv2 Multi-Loss Error Function

$$lossErr =$$
$$\lambda_{coord} \sum_{i=0}^{S^2} \sum_{j=0}^{B} L_{ij}^{obj} [(z_x^{ij} - \hat{z}_x^{l})^2 + (z_y^{ij} - \hat{z}_y^{l})^2 + (z_w^{ij} - \hat{z}_w^{l})^2 + (z_h^{ij} - \hat{z}_h^{l})^2]$$
$$+\lambda_{obj} \sum_{i=0}^{S^2} \sum_{j=0}^{B} L_{ij}^{obj} (C_{ij} - 1)^2$$
$$+\lambda_{noobj} \sum_{i=0}^{S^2} \sum_{j=0}^{B} L_{ij}^{noobj} (C_{ij} - 0)^2$$
$$+\lambda_{class} \sum_{i=0}^{S^2} L_{ij}^{obj} \sum_{c \in classes} (p_{ij}(c) - \hat{p}^{l}(c))^2$$

If a cell does not contain an object of interest, then $L_{ij}^{noobj} = 1$, and the multi-loss error function resolves to a simple squared error function: $\lambda_{noobj} \sum_{i=0}^{S^2} \sum_{j=0}^{B} (C_{ij} - 0)^2$ where C_{ij} represents the predicted probability of being an object for a prediction j in the cell i.

If a cell contains an object of interest, then $L_{ij}^{obj} = 1$, and the multi-loss error function drops $\lambda_{noobj} \sum_{i=0}^{S^2} \sum_{j=0}^{B} (C_{ij} - 0)^2$ and simplifies to the following:

$$\lambda_{coord} \sum_{i=0}^{S^2} \sum_{j=0}^{B} [(z_x^{ij} - \hat{z}_x^{l})^2 + (z_y^{ij} - \hat{z}_y^{l})^2 + (z_w^{ij} - \hat{z}_w^{l})^2 + (z_h^{ij} - \hat{z}_h^{l})^2]$$

$$+\lambda_{obj} \sum_{i=0}^{S^2} \sum_{j=0}^{B} (C_{ij} - 1)^2$$

$$+\lambda_{class} \sum_{i=0}^{S^2} \sum_{c \in classes}^{B} (p_{ij}(c) - \hat{p}^{l}(c))^2$$

where $z_x^{ij}, z_y^{ij}, z_w^{ij}, z_h^{ij}$ represent predicted coordinate offsets from the anchor box, and $\hat{z}_x^{l}, \hat{z}_y^{l}, \hat{z}_w^{l}, \hat{z}_h^{l}$ represent coordinate offsets of the labeled object bounding box for which the prediction ij is responsible.

Note: An anchor box provides a candidate set of starting points for bounding box predictions. This topic is discussed in more detail in a later section.

Only a single anchor box that contains the highest intersection over union (IOU) value with the true bounding box is retained as the prior of best fit and is represented by $\hat{z}_x^l, \hat{z}_y^l, \hat{z}_w^l, \hat{z}_h^l$ offsets.

Intersection over union (IOU) is calculated as shown in Figure 4.5.

Figure 4.5: Intersection over Union

$$IOU = \frac{AreaOfOverlap}{AreaOfUnion} = \frac{\Box}{\blacksquare}$$

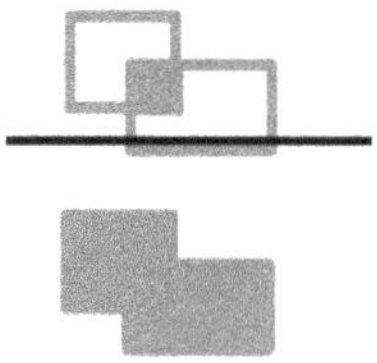

When a prediction is "responsible" for a true object, it means that the prediction is the closest match for a labeled object whose center falls into the cell, with respect to IOU. P_{ij} represents

the predicted probability of being class C for prediction j in the cell i. $\hat{P}^l(c)$ is 1 if the labeled object that this prediction is responsible for is class C. Otherwise, it is zero.

Error Function Scalars

Scalars ($\lambda_{coord}, \lambda_{obj}, \lambda_{noobj}, \lambda_{class}$) are used to fine-tune predictions by emphasizing specific predictions over others. For example, the authors of *You Only Look Once: Unified, Real-Time Object Detection* discovered that the prediction scores for confidence were being pressured toward zero because the number of cells not containing an object largely outnumbered the number of cells containing an object. Therefore, the authors used a lower scalar for λ_{noobj} (that is, 0.5) to offset the lack of objects in most grid cells. The authors further adjusted their predictions with a higher scalar for λ_{coord} (that is, 5) to upweight bounding box predictions (Redmon et al. 2016).

Changing scalars in the multi-loss error function imposes on the model a tradeoff between various perdition tasks. For example, increasing λ_{coord} causes the optimization process to emphasize the prediction of the bounding box at the sacrifice of de-emphasizing confidence and classification predictions. The λ_{coord} scalar is adjusted using the COORDSCALE= option in the object detection layer created by the last ADDLAYER statement.

Here is an example of COORDSCALE= implementation in the ADDLAYER statement:

```
AddLayer / model='ObjectDetection'
          name='output_object_dectect' layer={type='detection'
          coordScale = 5
 /***Remaining code not shown but will be displayed as topics are
introduced later in the chapter ***/
run;
```

Increasing λ_{obj} increases the model's awareness of objects at the cost of an increased false detection rate. The λ_{obj} scalar is adjusted using the OBJECTSCALE= option in the object detection layer created by the last ADDLAYER statement.

Here is an example of OBJECTSCALE= implementation in the ADDLAYER statement:

```
AddLayer / model='ObjectDetection'
          name='output_object_dectect' layer={type='detection'
          coordScale = 5
          objectScale = 1
 /***Remaining code not shown but will be displayed as topics are
introduced later in the chapter ***/
run;
```

Increasing λ_{noobj} encourages the model to be more cautious when detecting objects. Objects that exist in the picture are more likely to be missed if a large value is assigned to the λ_{noobj} scalar.

The λ_{noobj} scalar is adjusted using the PREDICTIONNOTAOBJECTSCALE= option in the object detection layer created by the last ADDLAYER statement.

Here is an example of PREDICTIONNOTAOBJECTSCALE= implementation in the ADDLAYER statement:

```
AddLayer / model='ObjectDetection'
          name='output_object_dectect' layer={type='detection'
          coordScale = 5
          objectScale = 1
          predictionNotAObjectScale = .25
 /***Remaining code not shown but will be displayed as topics are
introduced later in the chapter ***/
run;
```

Increasing λ_{class} encourages the model to pay more attention to correctly predicting the object classification, as opposed to the location of the object within the image. The λ_{class} scalar is adjusted using the CLASSSCALE= option in the object detection layer created by the last ADDLAYER statement.

Here is an example of CLASSSCALE= implementation in the ADDLAYER statement:

```
AddLayer / model='ObjectDetection'
        name='output_object_dectect' layer={type='detection'
        coordScale = 5
        objectScale = 1
        predictionNotAObjectScale = .25
        classScale = 1
 /***Remaining code not shown but will be displayed as topics are
introduced later in the chapter ***/
run;
```

> **Note:** CLASSSCALE= should be set to zero if the business need is to detect a single object. That is, no other objects are competing for the model's attention.

Anchor Boxes

An anchor box provides a candidate set of starting points for bounding box predictions. Including multiple anchor boxes of varying sizes decreases the need for the optimization process to adjust the bounding box coordinates dramatically because the chances of having a starting box that closely matches the box of an object increases. This means that there is a reduction in significant gradient swings and starting error is reduced.

Each anchor box is itself a candidate prediction. The candidate prediction (anchor box) must satisfy two criteria to become an actual prediction. First, the predicted probability of being an object for prediction in the cell i (C_{ij}) should be greater than or equal to the prediction threshold property value set by the DETECTIONTHRESHOLD= property.

Here is an example of DETECTIONTHRESHOLD= implementation in the ADDLAYER statement:

```
AddLayer / model='ObjectDetection'
        name='output_object_dectect' layer={type='detection'
        coordScale = 5
        objectScale = 1
        predictionNotAObjectScale = .25
        classScale = 1
        detectionThreshold = 0.15
 /***Remaining code not shown but will be displayed as topics are
introduced later in the chapter ***/
run;
```

The probability threshold value is set to 15% in the example code above. This means that a candidate prediction, j (anchor box), should have at least a predicted probability of 15% to be considered.

Candidate predictions that meet or exceed the posterior threshold are then assessed using the intersection over union (IOU) between the candidate prediction and the actual bounding box (ground truth). Candidate predictions that meet or exceed the IOU threshold set by the IOUTHRESHOLD= property become predictions.

Here is an example of IOUTHRESHOLD = implementation in the ADDLAYER statement:

```
AddLayer / model='ObjectDetection'
          name='output_object_dectect' layer={type='detection'
          coordScale = 5
          objectScale = 1
          predictionNotAObjectScale = .25
          classScale = 1
          detectionThreshold = 0.15
          iouThreshold = 0.1
 /***Remaining code not shown but will be displayed as topics are
introduced later in the chapter ***/
run;
```

The IOU threshold value is set to 10% in the example code above. This means that a candidate prediction, j (anchor box), should have at least an IOU value of 10% to become a prediction.

The appropriate number of anchor boxes and the size of the boxes is perhaps best determined empirically. Clustering algorithms applied to the training data guide the user to an appropriate number and size of anchor boxes. Prior sizes can be determined empirically through clustering on the pixel width and pixel height of actual bounding boxes. Average width and height for each cluster should be downsampled equivalently to that of the original image. For example, going from a 416 x 416 image to a 13 x 13 tensor means that the data has been downsampled by a factor of 32.

SAS software includes a rich set of clustering algorithms for grouping to the height and width of the labeled boxes. The average height and width of each group should be divided by the same amount for which the model downsampled the image. For example, a 416 x 416 image that has been downsampled to 13 x 13 has been decreased by a factor of 32. The ANCHORS= property specifies the number and size of the anchor boxes. Two numeric values are required for each anchor box. The first value represents the width of the anchor box, and the second value represents the height of the same box.

Here is an example of ANCHORS= implementation in the ADDLAYER statement:

```
AddLayer / model='ObjectDetection'
          name='output_object_dectect' layer={type='detection'
          coordScale = 5
          objectScale = 1
          predictionNotAObjectScale = .25
          classScale = 1
          detectionThreshold = 0.15
          iouThreshold = 0.1
          anchors ={10.71475, 10.43283,
               5.139861, 11.92218}
```

The code example above creates two anchor boxes with a square and rectangular shapes respectively.

Final Convolution Layer

There are some requirements for the output of the last layer of CNN, as well as the object detection layer parameters.

In the last convolutional layer, the width and heights of the output should both be equal to the value for gridNumber. The depth of the output should be the value calculated as predictionsPerGrid * (classNumber + coordNumber + 1).

The terms gridNumber, predictionsPerGrid, classNumber, and coordNumber are all parameters in the detection layer. The detection layer is the output layer of the object detection model.

SAS deep learning technologies can be accessed using many different types of code editors. SAS Studio is a code editor favorite among SAS users. SAS users can also code in the SAS language within Jupyter Notebook using a SAS kernel. R and Python programmers can also access SAS deep learning technologies through the Jupyter Notebook interface. DLPy is a high-level package for the Python APIs created for the deep learning technologies in SAS Viya. DLPy provides many prebuilt models, including VGG and ResNet. The pretrained weights using ImageNet data are also provided for those models. This would give you a warm start on your favorite task via transfer learning.

Demonstration: Using DLPy to Access SAS Deep Learning Technologies: Part 1

This demonstration uses DLPy to create and manipulate training image data sets for use with CNN-based object-detection models. The learning objective is to understand how you can use DLPy to create your own object detection training data set with only a few lines of code. You should also learn how to visually inspect an object detection data set for potential issues.

1. From the Windows menu, expand the **Anaconda3 (64-bit)** folder.
2. Select **Jupyter Notebook** to open the code editor interface, as seen in Figure 4.6.

Figure 4.6: Jupyter Notebook

3. Select **Documents** (Figure 4.7).

Figure 4.7: Documents

4. Select **SimpleObjectDetection.ipynb** (Figure 4.8).

Figure 4.8: jpynb Selection

5. Select **Cell ▶ Run All.**

Figure 4.9: Run All

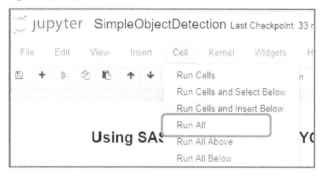

Demonstration: Using DLPy to Access SAS Deep Learning Technologies: Part 2

This demonstration examines the results from the object detection model that was run in the previous demo.

The example begins with configuring the computing environment. After that, the training data are prepared, followed by training the model using pre-trained weights for a warm start.

After you train the model, several holdout images are run against the trained model, and the results are visualized. Then several metrics (for example, precision) are generated to evaluate the model performance.

1. Navigate to Jupyter Notebook to view the code and results.
2. Begin by importing SWAT. SWAT is a Python interface to SAS Cloud Analytic Services. For more information about starting a CAS session with the SWAT package, see https://sassoftware.github.io/python-swat/getting-started.html.
3. Import DLPy and the functions for the DLPy utils, applications, and model classes.

```
import swat
import dlpy
from dlpy.utils import *
from dlpy.applications import *
from dlpy.model import *
import warnings
warnings.simplefilter(action='ignore', category=FutureWarning)
```

4. After configuring your environment and loading required libraries and functions, connect to your CAS server. You need a host name and port number for this step.

```
# CAS(host name,port,username,password)
s = swat.CAS("http://sasviya01.demo.sas.com/cas-shared-default-
http/", 8777, "student", "Metadata0", protocol="http")

s.table.addcaslib(activeonadd=False,
                  datasource={'srctype':'path'},
                  name='dnfs',
                  path='/home/student/LWDLUS/Image
Data/ObjectDetectionData',
                  subdirectories=False)
```

The results show that the caslib was successfully defined, as seen in Figure 4.10.

Figure 4.10: Successfully Defined caslib

```
NOTE: Cloud Analytic Services added the caslib 'dnfs'.
```

§ CASLibInfo

	Name	Type	Description	Path	Definition	Subdirs	Local	Active	Personal	Hidden	Transient
0	dnfs	PATH		/home/student/LWDLUS/Image Data/ObjectDetectio...		0.0	1.0	0.0	0.0	0.0	0.0

elapsed 0.000892s · user 0.000877s · mem 0.727MB

5. Prepare the data for training. This step accesses a large trained object detection data set saved as a SASHDAT file, creates a smaller subset of the training data, and then saves and loads the subsetted training data as a CAS table. The following code filters the source data by selecting only observations that contain single-object images classified as either 'Cat', 'Dog', or 'Bird'.

```
s.loadtable('train_3classes_1object_animals_cleaned.sashdat',
            caslib='dnfs',
            casout=dict(name='trainset', replace=1),
            where="_nObjects_ eq 1")
```

The output of the code cell is shown below in Figure 4.11.

Figure 4.11: Output of Code Cell

```
NOTE: Cloud Analytic Services made the file train_3classes_1object_animals_cleaned.sashdat available as table TRAINSET in casli
b CASUSER(student).

§ caslib
CASUSER(student)


§ tableName
TRAINSET


§ casTable
CASTable('TRAINSET', caslib='CASUSER(student)')

elapsed 2 31s  user 0 866s  sys 1 44s  mem 25 1MB
```

6. Examine the number of images corresponding to each class.
    ```
    s.simple.freq(table=dict(name='trainset', where='_istrain_ eq 1'),
    inputs='_object0_')
    ```
    The output of the code cell is shown in Figure 4.12.

Figure 4.12: Output of Code Cell

Frequency for TRAINSET

	Column	CharVar	FmtVar	Level	Frequency
0	_Object0_	Bird	Bird	1	6073.0
1	_Object0_	Cat	Cat	2	4606.0
2	_Object0_	Dog	Dog	3	5396.0

Examining the characteristics of the data shows the appropriately formatted target variables (Figure 4.13).
```
s.columnInfo(table='trainset')
```

Figure 4.13: Target Variables

	Column	ID	Type	RawLength	FormattedLength	Format	NFL	NFD
0	_id_	1	char	16	16		0	0
1	_image_	2	varbinary	170040	170040		0	0
2	_Object0_	3	varchar	4	27	$	27	0
3	_Object0_x	4	double	8	12		0	0
4	_Object0_width	5	double	8	12		0	0
5	_Object0_y	6	double	8	12		0	0
6	_Object0_height	7	double	8	12		0	0
7	_nObjects_	8	double	8	12		0	0
8	_IsTrain_	9	double	8	12		0	0

7. Now look at a sampling of the various anchor box shapes in the subsetted training data **trainset**.

```
yolo_anchors = get_anchors(s, data=s.CASTable(name='trainset',
where='_IsTrain_ eq 1'), n_anchors=5, coord_type='yolo')
yolo_anchors
```

Figure 4.14: Sampling of Anchor Box Shapes

```
(4.650501010599539,
 6.036841429943693,
 6.799603612467017,
 10.25466469920845,
 11.331232158567769,
 11.681879968286449,
 9.186354916042571,
 7.064349794639308,
 2.3086625863131673,
 2.7385131966707537)
```

8. Use DLPy to create a Tiny YOLOv2 object detection model named **yolo_model**. The model has three output classes and is configured to generate five predictions per grid in the detection layer of the model. The maximum is two boxes (and box labels) per image, and the table **yolo_anchors** contains the anchor shapes. See Figure 4.15.

```
yolo_model = Tiny_YoloV2(s,
                         n_classes=3,
                         predictions_per_grid=5,
                         anchors = yolo_anchors,
                         max_boxes=2,
                         coord_type='yolo',
                         max_label_per_image = 1,
                         class_scale=1.0,
                         coord_scale=1.0,
                         prediction_not_a_object_scale=1,
                         object_scale=5,
                         detection_threshold=0.2,
                         iou_threshold=0.2
```

Figure 4.15: Log

```
NOTE: Input layer added.
NOTE: Convolution layer added.
NOTE: Batch normalization layer added.
NOTE: Pooling layer added.
NOTE: Convolution layer added.
NOTE: Batch normalization layer added.
NOTE: Pooling layer added.
NOTE: Convolution layer added.
NOTE: Batch normalization layer added.
NOTE: Pooling layer added.
NOTE: Convolution layer added.
NOTE: Batch normalization layer added.
NOTE: Pooling layer added.
NOTE: Convolution layer added.
NOTE: Batch normalization layer added.
NOTE: Pooling layer added.
NOTE: Convolution layer added.
NOTE: Batch normalization layer added.
NOTE: Pooling layer added.
NOTE: Convolution layer added.
NOTE: Batch normalization layer added.
NOTE: Convolution layer added.
NOTE: Batch normalization layer added.
NOTE: Convolution layer added.
NOTE: Detection layer added.
NOTE: Model compiled successfully.
```

9. Use **print_summary** to view a table describing the model that we just created.

```
yolo_model.print_summary()
```

Figure 4.16: Print_summary Results

	Layer Id	Layer	Type	Kernel Size	Stride	Activation	Output Size	Number of Parameters
0	0	Input1	input			None	(416, 416, 3)	(0, 0)
1	1	Convo.1	convo	(3, 3)	1	Identity	(416, 416, 16)	(432, 0)
2	2	B.N.1	batchnorm			Leaky	(416, 416, 16)	(0, 32)
3	3	Pool1	pool	(2, 2)	2	Max	(208, 208, 16)	(0, 0)
4	4	Convo.2	convo	(3, 3)	1	Identity	(208, 208, 32)	(4608, 0)
5	5	B.N.2	batchnorm			Leaky	(208, 208, 32)	(0, 64)
6	6	Pool2	pool	(2, 2)	2	Max	(104, 104, 32)	(0, 0)
7	7	Convo.3	convo	(3, 3)	1	Identity	(104, 104, 64)	(18432, 0)
8	8	B.N.3	batchnorm			Leaky	(104, 104, 64)	(0, 128)
9	9	Pool3	pool	(2, 2)	2	Max	(52, 52, 64)	(0, 0)
10	10	Convo.4	convo	(3, 3)	1	Identity	(52, 52, 128)	(73728, 0)
11	11	B.N.4	batchnorm			Leaky	(52, 52, 128)	(0, 256)
12	12	Pool4	pool	(2, 2)	2	Max	(26, 26, 128)	(0, 0)
13	13	Convo.5	convo	(3, 3)	1	Identity	(26, 26, 256)	(294912, 0)
14	14	B.N.5	batchnorm			Leaky	(26, 26, 256)	(0, 512)
15	15	Pool5	pool	(2, 2)	2	Max	(13, 13, 256)	(0, 0)
16	16	Convo.6	convo	(3, 3)	1	Identity	(13, 13, 512)	(1179648, 0)
17	17	B.N.6	batchnorm			Leaky	(13, 13, 512)	(0, 1024)
18	18	Pool6	pool	(2, 2)	1	Max	(13, 13, 512)	(0, 0)
19	19	Convo.7	convo	(3, 3)	1	Identity	(13, 13, 1024)	(4718592, 0)
20	20	B.N.7	batchnorm			Leaky	(13, 13, 1024)	(0, 2048)
21	21	Convo.8	convo	(3, 3)	1	Identity	(13, 13, 512)	(4718592, 0)
22	22	B.N.8	batchnorm			Leaky	(13, 13, 512)	(0, 1024)
23	23	Convo.9	convo	(1, 1)	1	Identity	(13, 13, 40)	(20480, 0)
24	24	Detection1	detection			Auto	(13, 13, 40)	(0, 0)
25								11034512

Training an accurate object detection model can be time consuming and computationally expensive. Importing weights from an already trained YOLOv2 model saves time and can produce reasonable results.

```
s.table.loadtable(casout={'name':'tinyyolov2_40epoch_trainless100',
'replace':True},
                  caslib='dnfs',
                  path="tinyyolov2_40epoch_trainless100.sashdat")
```

Figure 4.17: Log

NOTE: Cloud Analytic Services made the file tinyyolov2_40epoch_trainless100.sashdat available as table TINYYOLOV2_40EPOCH_TRAIN
LESS100 in caslib CASUSER(student).

§ caslib
CASUSER(student)

§ tableName
TINYYOLOV2_40EPOCH_TRAINLESS100

§ casTable
CASTable('TINYYOLOV2_40EPOCH_TRAINLESS100', caslib='CASUSER(student)')

elapsed 0.00161s user 0.000252s sys 0.0013s mem 0.735MB

10. Create a table named **targets** and a table named **inputVars**. The **targets** table contains a list of objects along with their bounding box locations. The model hyperparameters are specified in the next code cell and the model is then trained for one epoch.

```
solver = MomentumSolver(learning_rate=0.0005, clip_grad_max = 100,
        clip_grad_min = -100)
        optimizer = Optimizer(algorithm=solver, mini_batch_size=64,
        log_level=3, max_epochs=1, reg_l2=0.005)
        data_specs = [DataSpec(type_='IMAGE', layer='Input1',
        data=inputVars),
        DataSpec(type_='OBJECTDETECTION', layer='Detection1',
                data=targets)]
        gpu = Gpu(devices=[0])
        yolo_model.set_weights('tinyyolov2_40epoch_trainless100')
```

Figure 4.18: Log

```
NOTE: Model weights attached successfully!
NOTE: Training based on existing weights.
NOTE: Only 1 out of 2 available GPU devices are used.
NOTE:  The Synchronous mode is enabled.
NOTE:  The total number of parameters is 11031952.
NOTE:  The approximate memory cost is 301.00 MB.
NOTE:  Loading weights cost      0.78 (s).
NOTE:  Initializing each layer cost     2.32 (s).
NOTE:  The total number of threads on each worker is 1.
NOTE:  The total mini-batch size per thread on each worker is 64.
NOTE:  The maximum mini-batch size across all workers for the synchronous mode is 64.
NOTE:  Epoch Learning Rate        Loss      IOU    Time(s)
NOTE:  0            0.001         13.72    0.5672   25.52
NOTE:  1            0.001         5.962    0.6527   24.93
NOTE:  2            0.001         5.588    0.6645   24.92
NOTE:  3            0.001          5.18    0.6711   24.73
NOTE:  4            0.001         4.997    0.6797   24.81
NOTE:  5            0.001         4.719    0.6885   25.00
NOTE:  6            0.001         4.655    0.7002   25.14
NOTE:  7            0.001         4.518    0.7107   25.21
NOTE:  8            0.001         4.325    0.7218   25.08
NOTE:  9            0.001         4.241    0.7268   25.12
NOTE:  The optimization reached the maximum number of epochs.
NOTE:  The total time is    250.47 (s).
```

11. Finally, use the trained model to score the holdout images and print the images with the model predictions. See Figure 4.19.

```
yolo_model.predict(data=s.CASTable(name='trainset',
    where='_istrain_ eq 0'), gpu = Gpu(devices=[0]))

    display_object_detections(conn=s,
                            coord_type='yolo',
                            max_objects=1,

                table=dict(name=yolo_model.valid_res_tbl.name,
                    where='_nobjects_ eq 1')
                    #,num_plot=10,
                    #n_col=3
                    )
```

Figure 4.19: Results.

Chapter 5: Computer Vision Case Study

This chapter details a computer vision case study using the SAS programming language.

One of the world's foremost medical research centers seeks to build a computer vision model to automatically extract and summarize an understanding of metabolic pathways from an image depicting the structure. That is, understanding the geometric sequence created by a series of chemical compounds and arrows. The images are parsed from research papers or journals.

This case study details a solution that involves creating a two-stage computer vision model. The first stage relies on an object localization (detection) model. An image classifier is used in the second stage to classify the orientation of an extracted arrow. **However, only the first stage is described in this case study.**

The training data consist of 871 labeled images, of which approximately half of the images were extracted from papers and the other half of the images were synthetically generated by hand. The synthetic images have variants of arrows positioned in random orientation in combination with randomly selected compounds. Each synthetic image contains noise to improve generalization (text, shapes, random noise patches, and so on). Additionally, the synthetic images were designed so that each arrow and compound could be "boxed" without capturing other protruding edges in hopes of reducing entity obfuscation, as seen in Figure 5.1.

Figure 5.1: Synthetic Images

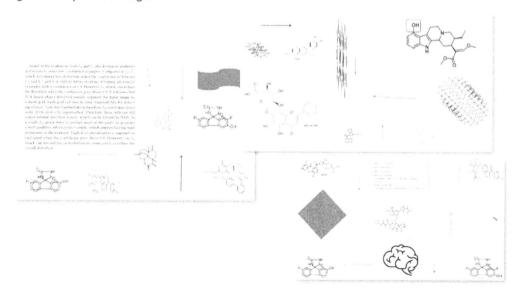

An example of several labeled images can be seen in Figure 5.2.

Figure 5.2: Labeled Images

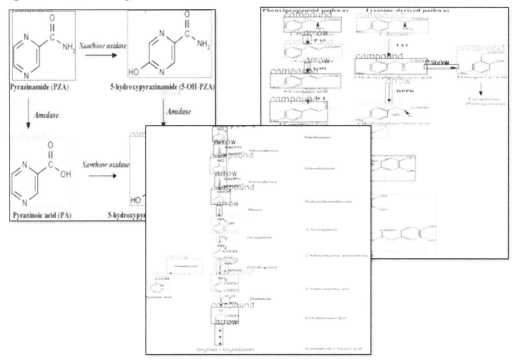

The data is in a SASHDAT data file titled **Train_metabolic_pathways**.

1. Open the program titled CASESTUDY_1a. This program begins with a macro variable defining the location of the data and programs.

    %let datalocation= *specify the location of the programs and data here;

    Then the cas libraries are established.

    ```
    libname mycas cas;
    proc cas;
        table.addCaslib / name='Scoreimageslib'
    path="&datalocation"
                    subdirectories=true;
    quit;
    ```

2. Next, a macro is created that is used to create a list of output variable names corresponding to the maximum number of objects in an image that exist in the training data. In this case, the largest number of objects in an image is 41. The list of output variables includes the number of objects in the image (_nobjects_) and a set of variables defining the spatial location of each entity. (ex. _Object#_, _Object#_x, _Object#_y, _Object#_width, and _Object#_height). The list of variable names is stored in a macro variable called **ObjectTargets**.

    ```
    %macro makevars;
    data mycas.createvars;
    _nObjects_=.;
    %do i=0 %to 40;
    ```

```
length _Object&i._ varchar(8);
length _Object&i._x _Object&i._y _Object&i._width
_Object&i._height 8;
_Object&i._='        ';
_Object&i._x=.;
_Object&i._y=.;
_Object&i._width=.;
_Object&i._height=.;
%end;
run;
%mend makevars;
%makevars;

 proc transpose data=mycas.createvars
out=createvars_Transposed;
 var _all_;
 run;
 proc sql;
  select quote(trim(_NAME_), "'")
         into: ObjectTargets separated by ", "
         from work.createvars_Transposed;
  quit;
```

3. Next, the program uploads the Train_metabolic_pathways data.

```
proc casutil;
load file="&datalocation/Train_metabolic_pathways.sashdat"
/*TrainsetFull.sashdat    TrainsetFullNew*/
casout="ChemArrows"
importoptions=(filetype="hdat")
replace;
quit;
```

The images in the training data have been resized to 416 x 416 and have been converted to grayscale. The data also includes a partition indicator, _partind_.

4. Next, the program loads a pretrained model called Tiny-Yolov2. SAS provides a large number of pretrained models that can be retrieved from the following URL: https://support.sas.com/documentation/prod-p/vdmml/zip/index.html.

```
/* Load model - TinyYolov2 */
proc casutil;
load file="&datalocation/Tiny-Yolov2.sashdat"
casout="Tiny-Yolov2"
importoptions=(filetype="hdat")
replace;
quit;

proc casutil;
load file="/data/roblan/Models/Pretrained/Tiny_Yolo/Tiny-
Yolov2_weights.sashdat"
casout="Tiny-Yolov2_weights"
importoptions=(filetype="hdat")
replace;
quit;
```

5. The model table is displayed using the fetch table action. Several action sets are also loaded.

```
proc cas;
  table.fetch  /
   table="Tiny-Yolov2"
   to=500;
  loadactionset 'deeplearn';
  loadactionset 'image';
  loadactionset 'table';
quit;
```

Figure 5.3 shows the results.

Figure 5.3: Model Table

Selected Rows from Table TINY-YOLOV2

Index	_DLKey0_	_DLKey1_	_DLChrVal_	_DLNumVal_	_DLLayerID_
1	input1	inputopts.nchannels	nchannels	3	0
2	convo.3	convoopts.height	height	3	7
3	convo.7	srclayers.0	pool6	18	19
4	convo.7	convoopts.std	std	1	19
5	convo.5	convoopts.act	Identity	1	13
6	convo.4	convoopts.mean	mean	0	10
7	detection1	detectionopts.jitter	jitter	0.2	24
8	convo.4	layertype	Convolution Layer	2	10
9	pool1	poolingopts.poolingtype	Max Pooling	1	3
10	convo.4	convoopts.init	Xavier	1	10
11	convo.2	convoopts.pad_top	pad_top	-1	4
12	convo.1	convoopts.act	Identity	1	1
13	convo.9	convoopts.mean	mean	0	23
14	b.n.1	srclayers.0	convo.1	1	2
15	convo.9	srclayers.0	b.n.8	22	23
16	convo.2	convoopts.pad_left	pad_left	-1	4
17	b.n.7	srclayers.0	convo.7	19	20
18	pool6	srclayers.0	b.n.6	17	18
19	convo.1	convoopts.std	std	1	1
...					
328	pool4	poolingopts.poolingtype	Max Pooling	1	12
329	convo.9	convoopts.stride_height	stride_height	1	23
330	convo.8	convoopts.init	Xavier	1	21
331	pool2	poolingopts.width	width	2	6
332	detection1	detectionopts.anchors.2	anchors	2.98513005	24
333	pool4	srclayers.0	b.n.4	11	12
334	convo.2	srclayers.0	pool1	3	4

The model table defines the architecture of the model and can be altered to change or enhance the behavior of the pretrained model. For example, the pretrained Tiny-Yolov2 provided by SAS does not perform any in-memory flipping mutations of input images. This behavior is modified in subsequent code enabling the model to flip the incoming images. The coordinates are automatically adjusted to correct for the geometric mutations.

The output layer (detections1) is removed from the Tiny-Yolov2 model.

```
proc cas;
removelayer / model='Tiny-Yolov2' name='detection1';
quit;
```

Several adaptation layers are added along with a new output layer.

```
proc cas;


AddLayer / model='Tiny-Yolov2' name='ConVL1'
layer={type='CONVO' nFilters=1000 /*588*/ width=1 height=1
stride=1 act='identity' includeBias=FALSE}
srcLayers={'convo.9'};
AddLayer / model='Tiny-Yolov2' name='BatchL1'
layer={type='BATCHNORM' act='ELU'} srcLayers={'ConVL1'};

AddLayer / model='Tiny-Yolov2' name='ConVL2'
layer={type='CONVO' nFilters=1000 /*588*/ width=3 height=3
stride=1 act='identity' includeBias=FALSE}
srcLayers={'BatchL1'};
AddLayer / model='Tiny-Yolov2' name='BatchL2'
layer={type='BATCHNORM' act='ELU'} srcLayers={'ConVL2'};

AddLayer / model='Tiny-Yolov2' name='PoolL1'
layer={type='POOL'  width=2 height=2 stride=1 pool='max'}
srcLayers={'BatchL2'};

AddLayer / model='Tiny-Yolov2' name='ConVL3a'
layer={type='CONVO' nFilters=250 /*45*/ width=1 height=1
stride=1 act='identity' includeBias=FALSE}
srcLayers={'PoolL1'};
AddLayer / model='Tiny-Yolov2' name='BatchL3a'
layer={type='BATCHNORM' act='ELU'} srcLayers={'ConVL3a'};


AddLayer / model='Tiny-Yolov2' name='ConVL3'
layer={type='CONVO' nFilters=250 /*2000*/ width=1 height=1
stride=1 act='identity' includeBias=FALSE}
srcLayers={'PoolL1'};
AddLayer / model='Tiny-Yolov2' name='BatchL3'
layer={type='BATCHNORM' act='ELU'} srcLayers={'ConVL3'};

AddLayer / model='Tiny-Yolov2' name='ConVL4'
layer={type='CONVO' nFilters=250 /*2000*/ width=3 height=3
stride=1 act='identity' includeBias=FALSE}
srcLayers={'BatchL3'};
AddLayer / model='Tiny-Yolov2' name='BatchL4'
layer={type='BATCHNORM' act='ELU'} srcLayers={'ConVL4'};

AddLayer / model='Tiny-Yolov2' name='ConVL5a'
layer={type='CONVO' nFilters=25 /*250*/ width=1 height=1
stride=1 act='identity' includeBias=FALSE}
srcLayers={'BatchL4'};
AddLayer / model='Tiny-Yolov2' name='BatchL5a'
layer={type='BATCHNORM' act='ELU'} srcLayers={'ConVL5a'};
```

```
AddLayer / model='Tiny-Yolov2' name='ConVL5'
layer={type='CONVO' nFilters=25 /*250*/  width=3 height=3
stride=1 act='identity' includeBias=FALSE}
srcLayers={'BatchL5a'};

AddLayer / model='Tiny-Yolov2' name='BatchL5'
layer={type='BATCHNORM' act='ELU'} srcLayers={'ConVL5'};

AddLayer / model='Tiny-Yolov2' name='concatl1'
layer={type='concat'} srcLayers={'BatchL5','BatchL3a'};

AddLayer / model='Tiny-Yolov2' name='ConVL7'
layer={type='CONVO' nFilters=49  width=1 height=1 stride=1
act='identity' includeBias=FALSE} srcLayers={'concatl1'};

AddLayer / model='Tiny-Yolov2'
name='detection1' layer={type='detection'

 detectionModelType = "YOLOV2"

 classNumber = 2

 gridNumber = 13

 coordNumber = 4

 predictionsPerGrid = 7

 anchors ={
0.459726,
1.103842,
4.418608,
2.111627,
1.453295,
2.195889,
1.318375,
0.840207,
1.912652,
5.398259,
0.485043,
0.356428,
2.295685,
1.433674
}
                                    softMaxForClassProb =
True
                                       objectScale = 1.01

 predictionNotAObjectScale =1
                                       classScale = 1.0
                                       coordScale = 3
                                       coordType = "YOLO"
                                       detectionThreshold
= 0.4
                                       iouThreshold = 0.1
trainIouThreshold=.3
/*                                     randomBoxes=TRUE
*/
```

```
                                                  }
            srcLayers={'ConVL7'}; quit;
```

6. The pretrained model architecture is further modified with the following characteristics using the update table action:

   ○ Images are randomly selected and flipped horizontally

   ○ Images are randomly selected and flipped vertically

   ○ A dropout rate of 0.0176666749 is applied to all layers except the input and first convolution layers

Changing the _DLNumVal_ variable's value to "4" for the input layer's flip mutation option tells SAS to vertically and horizontally flip randomly selected images. Other modifications such as random cropping and dropout are also applied to the model.

```
proc cas;
/* Apply horizontal and vertical flipping mutations */
mytbl.name  ="Tiny-Yolov2";
mytbl.where = "'input1' = _DLKey0_ and 'No flipping' =
_DLChrVal_";
table.update /
    table=mytbl
    set = {
      {var="_DLNumVal_", value="4"}};
/* Apply a random cropping mutation */
mytbl.name  ="Tiny-Yolov2";
mytbl.where = "'input1' = _DLKey0_ and 'No cropping' =
_DLChrVal_";
table.update /

    table=mytbl
    set = {
      {var="_DLNumVal_", value="2"}

    };
/* Apply a dropout rate of 0.0176666749 to all layers in the
model */
mytbl.name  ="Tiny-Yolov2";
mytbl.where = "'dropout' = _DLChrVal_";
table.update /
    table=mytbl
    set = {
      {var="_DLNumVal_", value="0.0176666749"}};
/* Reset the dropout rate for the first convolution layer
back to zero */
mytbl.name  ="Tiny-Yolov2";
mytbl.where = "'convo.1' = _DLKey0_ and 'dropout' =
_DLChrVal_";
table.update /
    table=mytbl
    set = {
      {var="_DLNumVal_", value="0"}};

/* Reset the dropout rate for the input layer back to zero */
mytbl.name  ="Tiny-Yolov2";
```

```
mytbl.where = "'input1' = _DLKey0_ and 'dropout' =
_DLChrVal_";
table.update /
    table=mytbl
    set = {
      {var="_DLNumVal_", value="0.00003"}};    quit;
```

The Tiny-Yolov2 model was previously trained using DLTRAIN for 250 epochs and the loss for each partition plotted, as shown in Figure 5.4.

Figure 5.4: Tiny-Yolov2 Model, DLTRAIN 250 Epochs with Losses Partition Plotted

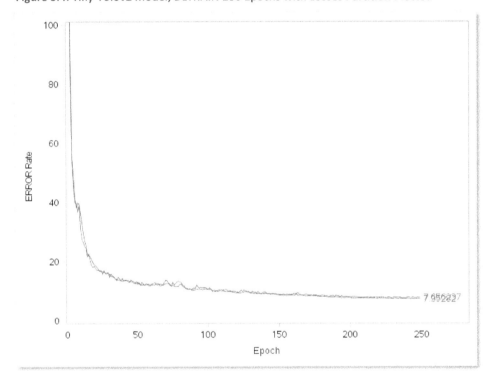

7. We continue to train the model for an additional epoch.

```
proc cas;
  dlTrain / table={name='Train_metabolic_pathways',
where='_PartInd_=1'} model='Tiny-Yolov2'
        modelWeights={name='ConVTrainedWeights_d', replace=1}
        bestweights={name='MytinyYolo', replace=1}
initWeights='tinyYolo_pathways'
dataSpecs={{data={&ObjectTargets},
                                layer='detection1',
                           type='OBJECTDETECTION'}

                      {data={'_image_'},
                         layer='input1',
                         type='IMAGE'}
                  }        GPU=True
forceEqualPadding = True
```

```
    ValidTable={name='Train_metabolic_pathways',
 where='_PartInd_=2'}
        optimizer={minibatchsize=4, ignoreTrainingError=true,
            algorithm={method='LARS', lrpolicy='Step',
 gamma=0.901, stepsize=1,       learningrate=6.047E-7,
 scalefactor=.009,warmup=0,momentum=0.9395666667,
 clipgradmin=-100, clipgradmax=100},loglevel=3,
 regL1=0.0022902367, regL2=0.0025301567,maxepochs=1}
 seed=12345
 ;
 quit;
```

The model contains 23,511,562 parameters! Performance is marginally worse than where the model landed during the extended training session.

8. The trained model is used to predict (score or inference) the locations of compounds and arrows from the data.

```
    /* Score full data to generate extractions */;
 proc cas;
            dlScore /
 table={name='Train_metabolic_pathways'} model='Tiny-Yolov2'

 initWeights='tinyYolo'

 casout={name='PathwaysScored', replace=1}

 copyVars={'_image_'}

 gpu=true;
 Quit;
```

9. The predictions from DLSCORE are passed to the extractDetectedObjects action. The extractDetectedObjects action applies the predicted bounding boxes to the images in the data, PathwaysScored.

```
 proc cas;
 image.extractDetectedObjects /
   casOut={name='ObjectsExtracted', replace=true}
   coordType='YOLO'
   maxObjects=50
        extractType='highlight'
   Table={name='PathwaysScored'};
 quit;
```

10. The images in PathwaysScored are saved and viewed.

```
 proc cas;
    saveImages / caslib="Scoreimageslib"
    subdirectory='Predicted Locations'
    images = {table={name='ObjectsExtracted'} image='_image_'}
    overwrite=TRUE    type="JPG";

  image.loadimages / caslib='Scoreimageslib' path='Predicted
 Locations'
                                      recurse=true
 casout={name='Imagesforviewing', replace=true};

 quit;
```

```
/***************/
/* View Images */
/***************/

data _null_;
  set mycas.Imagesforviewing(where=(
  _id_<=6)
        keep=_path_ _id_ ) end=eof;
  if _n_=1 then
        do;
              dcl odsout obj();
              obj.layout_gridded(columns:1);
        end;
  obj.region();
  obj.format_text(just: "c", style_attr: 'font_size=8pt');
  obj.image(file: _path_, width: "416", height: "416");

  if eof then
        do;
              obj.layout_end();
        end;
run;
```

Figures 6.5 and 6.6 show the results.

Figure 5.5: Results

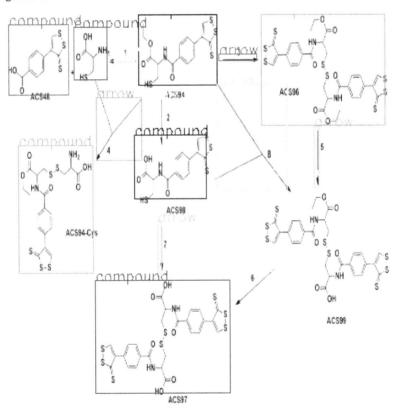

Figure 5.6: Results

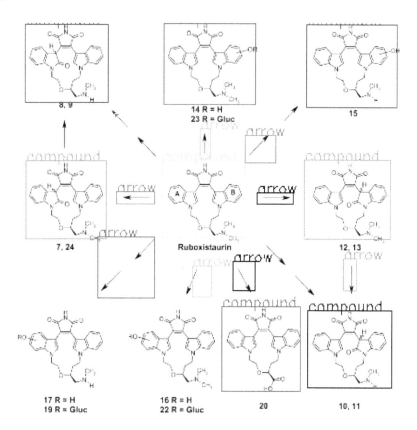

References

Bengio, Y., J. Louradour, R. Collobert, and J. Weston. 2009. "Curriculum Learning." Proceedings of the 26th International Conference on Machine Learning. Montreal, Canada. pp. 41–48.

Daniely, A., R. Frostig, and Y. Singerz. 2017. "Toward Deeper Understanding of Neural Networks: The Power of Initialization and a Dual View on Expressivity." arXiv:1602.05897v2 [cs.LG]. Cornell University Library. New York. Available at: https://arxiv.org/pdf/1602.05897v1.pdf.

Glorot, X., A. Bordes, and Y. Bengio. 2011. "Domain Adaption for Large-Scale Sentiment Classification: A Deep Learning Approach." Available at: http://www.icml-2011.org/papers/342_icmlpaper.pdf.

Goodfellow, I., Y. Bengio, and A. Courville. 2017. *Deep Learning*. Cambridge. MA: The MIT Press.

He, K., et al. 2015. "Deep Residual Learning for Image Recognition." ArXiv:1512.03385 [cs.CV] Cornell University Library. New York. Available at: https://arxiv.org/pdf/1512.03385.pdf.

Hertz, J., A. Krogh, and R. G. Palmer. 1991. *Introduction to the Theory of Neural Computation*. Redwood City, CA: Addison-Wesley Publishing Co.

Hinton, G. E., S. Osindero, and Y. W. Teh. 2006. "A Fast Learning Algorithm for Deep Belief Networks." *Neural Computation* 18:1527–1554.

Ioffe, S. and C. Szegedy. 2015. "Batch Normalization: Accelerating Deep Network Training by Reducing Internal Covariate Shift." arXiv:1502.03167v3 [cs.LG]. Cornell University Library. New York. Available at: https://arxiv.org/abs/1502.03167.

Kingma, D. and J. Lei Ba. 2017. "ADAM: A METHOD FOR STOCHASTIC OPTIMIZATION." arXiv:1412.6980v9 [cs.LG]. Cornell University Library. New York. Available at: https://arxiv.org/abs/1412.6980.

Koch, P., et al. 2018. "Autotune: A Derivative-free Optimization Framework for Hyperparameter Tuning." arXiv:1804.07824v2 [cs.LG]. Cornell University Library. New York. Available at https://arxiv.org/abs/1804.07824.

Li, L. et al. 2018. "Hyperband: A Novel Bandit-Based Approach to Hyperparameter Optimization." arXiv:1603.06560v4 [cs.LG]. Cornell University Library. New York. Available at: https://arxiv.org/abs/1603.06560.

McCulloch, W. and W. Pitts. 1943. *A Logical Calculus of the Ideas Immanent in Nervous Activity*. Pergamon Press plc. Society for Mathematical Biology

Ng, A. 2013. "Stochastic Gradient Descent" Video from Coursera - Stanford University - Course: Machine Learning: Published on Nov 1, 2013. https://www.coursera.org/course/ml.

Principe, J.C., N. R. Euliano, and W. C. Lefebvre. 2000. *Neural and Adaptive Systems*. New York: Wiley.

Redmon, J., et al. 2016. "You Only Look Once: Unified, Real-Time Object Detection." arXiv:1506.02640v5 [cs.CV]. Cornell University Library. New York. Available at https://arxiv.org/abs/1506.02640.

Saining, X., et al. 2017. "Aggregated Residual Transformations for Deep Neural Networks" arXiv:1611.05431v2 [cs.CV] Cornell University Library. New York. Available at https://arxiv.org/abs/1611.05431.

Santurkar, S., et al. 2018. "How Does Batch Normalization Help Optimization? (No, It Is Not About Internal Covariate Shift)." arXiv:1805.11604v2 [stat.ML]. Cornell University Library. New York. Available at https://arxiv.org/abs/1805.11604.

Springenberg, J., et al. "Striving for Simplicity: The All Convolutional Net." Available at: https://arxiv.org/abs/1412.6806.

Szegedy, C., et al. 2014. "Going Deeper with Convolutions." arXiv:1409.4842v1 [cs.CV]. Cornell University Library. New York. Available at https://arxiv.org/abs/1409.4842.

Vanhoucke, V., Senior, A., and Mao, M. Z. 2011. Improving the speed of neural networks on CPUs. In: *Proc. Deep Learning and Unsupervised Feature Learning NIPS Workshop*.

Vincent, P., et al. 2008. "Extracting and Composing Robust Features with Denoising Autoencoders." Montreal: Université de Montreal.

Vincent, P., et al. 2010. "Stacked Denoising Autoencoders: Learning Useful Representations in a Deep Network with a Logical Denoising Criterion." *Journal of Machine Learning Research* 11:3371–3408.

Weiss, K., T. Khoshgoftaar, and D. Wang. 2016. "A Survey of Transfer Learning." *Journal of Big Data* 3: Article number: 9

Ready to take your SAS® and JMP® skills up a notch?

Be among the first to know about new books,
special events, and exclusive discounts.
support.sas.com/newbooks

Share your expertise. Write a book with SAS.
support.sas.com/publish

www.ingramcontent.com/pod-product-compliance
Lightning Source LLC
Chambersburg PA
CBHW080535060326
40690CB00022B/5139